Cathodoluminescence
and Photoluminescence

Theories and Practical Applications

Phosphor Science and Engineering

Series Editor

William M. Yen

University of Georgia
Athens, Georgia

Cathodoluminescence and Photoluminescence

Theories and Practical Applications

Lyuji Ozawa

CRC Press
Taylor & Francis Group
Boca Raton   London   New York

CRC Press is an imprint of the
Taylor & Francis Group, an **informa** business

CRC Press
Taylor & Francis Group
6000 Broken Sound Parkway NW, Suite 300
Boca Raton, FL 33487-2742

First issued in paperback 2019

ISBN-13: 978-1-4200-5270-1 (hbk)
ISBN-13: 978-0-367-38913-0 (pbk)

Library of Congress Cataloging-in-Publication Data

Ozawa, Lyuji.
    Cathodoluminescence and Photoluminescence : theories and practical applications / Lyuji Ozawa.
        p. cm.
    Includes bibliographical references and index.
    ISBN-13: 978-1-4200-5270-1 (alk. paper)
    ISBN-10: 1-4200-5270-5 (alk. paper)
    1. Phosphors. 2. Cathodoluminescence. 3. Fluorescent screens. I. Title.

QC479.5.O934 1989
621.3815'422--dc22   2006037695

**Visit the Taylor & Francis Web site at
http://www.taylorandfrancis.com**

**and the CRC Press Web site at
http://www.crcpress.com**

Table of Contents

Preface

Phosphor screens are currently in wide use as display devices and light sources. Display devices serve as the interface between information stored in electronics devices and humans. Light sources extend life activities from dark to comfortably illuminated rooms. The requirements for reliable display devices include (1) optimization of light output from phosphor screens, (2) long operational lifetime, and (3) low production cost. This book provides information for further study on the subject.

Phosphor screens are merely transducers — from energy of invisible particles to photons of visible wavelength. The luminescence generated by irradiation of electrons is called cathodoluminescence (CL), and luminescence generated by photons is called photoluminescence (PL). Practical phosphor screens of CL and PL are constructed by arranging a large number of tiny crystallized particles (4 μm; 10^6 particles per square centimeter). Notwithstanding that the production of phosphor powders and the screening of phosphor powders have a long history (more than 50 years), and there are numerous publications on both CL and PL, I have received many inquiries regarding CL and PL from both scientists and engineers who wish to improve display devices and light sources.

Many publications describe the formation of luminescence centers. The formation of luminescence centers in commercial phosphor particles is influenced by short-range perfection in volume of 100Å diameter), which allow for mass production of phosphor particles possessing irregular shapes and sizes. The energy conversion efficiencies of CL and PL were optimized theoretically and practically some 30 years ago. Other questions relate to the number of photons from phosphor screens and the lifetime of phosphor screens. It is realized that the main problems arise from the production technology of phosphor powders and the screening of phosphor powders on the substrate. Commercial phosphor powders are heavily contaminated by (1) residuals of the by-products, (2) deliberately adhered materials on the surface of phosphor particles, and (3) an interface layer of host crystal and by-products. The ideal phosphor screens are made from phosphor powder particles that have a clean surface, an irregular shape, and a narrow distribution with log-normal probability. Spherical particles are not acceptable in practice. The novel production process of phosphor powders, by the revised process of phosphor production for 50 years, allows the mass production of 1000 kg phosphor powder per lot. Additional details on CL and PL from phosphor screens are given quantitatively in this book, based on a basic knowledge of solid-state physics and materials science.

This book is intended for both scientists and engineers — especially younger scientists and engineers — who wish to further improve display devices and light sources that use PL and CL.

Lyuji Ozawa

Author

Lyuji Ozawa has been involved in production, engineering, and research activity on luminescent materials for more than 50 years — from basic research to applications of phosphor powders. His first work involved the determination and distribution of phosphor particles, in order to theoretically optimize the structure of phosphor screens. In Dr. Ozawa's distinguished carrer, he has made seminal and continuing contributions to our understanding of CRT display technologies and has been a leading advocate of improved phosphor material efficiencies, Hg-free flat fluourescent lamps and PDP.

Ozawa's current interest lies in the development of fluorescent lamps without mercury. He is the author of three books, a co-author of one book, and the author of many journal articles. He holds eight U.S. patents.

1

Introduction

Living standards have been markedly improved by illumination in the dark and communication skill by means of recording and reading information on various media. Illumination by light has evolved from wood fires, to torch flames, the burning of oil, candles, gas flames, incandescent lamps, fluorescent lamps, and light-emitting diodes (LEDs), combined with photoluminescence (PL) from phosphor particles. By improving illumination, life's activities are significantly prolonged into the night hours, especially by application of PL light from phosphor screens. Life's activities are supported by communication with others. Communication of information has evolved from the faces of rock cliffs, to the walls of cave, clay tablets, parchment, wood and bamboo plates, sheets of paper, magnetic tapes (and disks), and electronic chips. Electronic devices have significantly increased the speed of the communication of information. Information stored on tapes and chips in electronic devices (e.g., TV sets and computers) are invisible to the human eye. Display devices have been developed as an interface between human and electronic devices for the visualization of invisible information in electronic devices. The images on display devices are illustrated on screens by cathodoluminescence (CL) or photoluminescence (PL) from phosphor screens. A good understanding of the generation of CL and PL from phosphor screens becomes an important subject in modern-day activities. This book discusses in detail the optimization of CL and PL generation in the tiny phosphor particles (4 to 5 μm) that comprise the phosphor screen. Notwithstanding, we thoroughly discuss CL in an effort to clarify the ambiguities in CL generation and practices; the results are directly applicable to practical PL.

Since Karl Ferdinand Braun invented the cathode ray tube (CRT) in 1897 [1], the CRT has remained a typical display device. Recently, liquid crystal displays (LCDs) and plasma display panels (PDPs), which use PL, were developed for the same purpose [2]. The CRT is dominant among the various display devices, with advantages that include high image luminance, wide viewing angle, and low total cost over the device lifespan, including saving energy. In CRT display devices, processed information is visualized by light from the CL from a phosphor screen, which arranges tiny particles (~4 μm) in thicknesses of a few layers on average. The particles in a phosphor screen are called *phosphors*. Under irradiation by an electron beam on the phosphor screen, phosphor particles emit CL. Light images are illustrated on the entire phosphor screen (e.g., TV and computer images) by scanning the electron beam on the phosphor screen from left to right and top to bottom within a frame cycle ($\geq$30 Hz). Although CL images on a phosphor screen illustrated by rapidly moving CL light spots (1 mm in diameter with a speed of 10^5 cm s^{-1}), one can perceive planar images on the entire screen (like printed images on sheets of paper) by the effects of afterimages on the human eye.

Many scientists and engineers have been involved in improving CRTs during the past 100 years. Their efforts achieved increased screen luminance (330 cd m^{-2}) and operational lifespans ($\geq$10 years) with low production costs. Consequently, TV viewers can watch CL images on phosphor screens in the comfort of their homes [3]. Families in industrialized countries usually have multiple TV sets and personal computers in their homes. The annual world-wide production volume reached 200 million CRTs in 2003, and the volume increases annually. It is said that CRT production technology is a "mature technology" that is, it is state of the art. Nevertheless, there is the long history and huge production volume, a further improvement of image quality on CRT screen requires in CRT producers, especially at a distance of distinct vision (~30 cm away from the screen) in competition with flat panel displays (FPDs) [2], such as LCD, PDP, electroluminescence devices (ELDs), and organic electroluminescence devices (OLEDs). FPDs do not exhibit flicker or smeared images screens at the distance of distinct vision. The images on CRT phosphor screens exhibit a small amount of flicker and smeared images, especially for high-light images. The human eye is permanently damaged by the observation of flicker and smeared images over long periods of time. This damage to the human eye mandates the substitution of CRTs by FPDs, although FPDs have high production costs, short operational lifetimes, and they are power-hungry devices.

CRT-TV sets have been accepted for a long time as long as the CL images are observed at a certain distance from the screen (i.e., three to five times the size of the screen). At that distance, the displacement of flicker and smeared images on phosphor screens decreases the resolution of the eye, which is given by the critical vision angle (3×10^{-4} radians) of the eyes, such that one does not perceive the flicker and smeared images. In that time, the primary concern of CL images were screen luminance that gives the equivalent images of daytime scenes (~1×10^{21} photons s^{-1} cm^{-2}) [4]. The phosphor screen merely acts as an energy transducer. The energy conversion efficiencies of the phosphors (output energy of emitted photons/input energy to the phosphor screen) were empirically optimized by phosphor manufacturers prior to 1970 [4]. That is, CL generation in phosphor particles has been well optimized empir-ically. The values of blue and green phosphors are 20%; those of white and red phosphors are 8%. Before 1970, the screen luminance from phosphor screens of color CRTs was, however, a low level (25 cd m^{-2}), which allowed the observation of TV images in a dim room (e.g., movie theater). The screen luminance of phosphor screens has improved more than ten times (350 cd m^{-2})

by manipulating CL properties, keeping the values constant [3, 4]; these include (1) expanding the phosphor pixel size in the phosphor screen by the application of a black matrix, (2) increasing the lumen-weight by a color shift of red CL using colorimetry, (3) increasing the anode voltages (from 15 to 25 kV), and (4) enlarging the phosphor particle sizes (from 2 to 4 μm). Each phosphor pixel in the CRT screen, which has a screen luminance of 350 cd m^{-2}, instantaneously emits photons equivalent to 1×10^{21} photons s^{-1} cm^{-2} [4]. This is why CRTs have used CL phosphor screens for more than half a century. Currently, CL images on phosphor screens of CRTs are observed at the distance of distinctive vision (~30 cm away from the screen). Then, the size of the flicker of CL phosphor screens becomes larger than the critical vision angle (3×10^{-4} radians) [3]; and thus one perceives flicker and smeared images. The complete removal of flicker and smeared images from CL screens is an urgent task of the CRT industry.

The practical CL phosphors are empirically selected as the brighter phosphors. They include cub-ZnS:Ag:Cl (blue), cub-ZnS:Cu:Al (green), and Y_2O_2S:Eu (red) in color CRTs, and Y_2O_2S:Eu:Tb (white) in black-and-white monochrome CRTs [3]. We may use those phosphor screens in future CRTs.

The image quality of phosphor screens in CRTs is predominantly determined by the optical and electrical properties of the bulk phosphor particles in the screen. These properties correlate with irradiation conditions of the electron beam on the phosphor screen, and do not relate to the CL properties generated in the phosphor particles. Although there are many books and review articles devoted to CL phosphors [3–18], there are a limited number of reports on the subject. A difficulty regarding scientific study comes from the presence of defects in phosphor screens, which are discontinuous media (softly piled particles). Furthermore, the phosphor powders are seriously contaminated with the residuals of phosphor production [19], and the surface of each phosphor particle is heavily contaminated with microclusters (insulators), such as SiO_2 (and pigments) to control the screening of phosphor powders on CRT faceplates.

Commercial phosphor powders usually contain some amount of strongly clumped (or bound) particles that generate the defects (pinholes and clumped particles) in the phosphor screen. For commercial phosphor powders, the pH value and the electric conductivity of the PVA (polyvinyl alcohol) phosphor slurry change with the stirring time of the screening slurry. The photosensitivity of dried PVA resin film, covering the phosphor particle, markedly changes with pH. The change in pH value of a PVA phosphor slurry has been practically solved by applying a 1-day aging process. Clumps of phosphor particles change with pH and electric conductivity in slurry. A clean surface of phosphor particles is essential for the screening of CL phosphor powders on faceplates [20]. The residuals at the contact gaps between phosphor particles dissolve in water, which slowly diffuses into the gaps. The diffusion of water into the gaps is a function of time and temperature. If commercial phosphor powders are soaked overnight in heated water at 80°C, the residuals in the gaps dissolve in the heated water [21]. There are phosphor powders that do not contain the residuals in these gaps, and are stable in both air and water. The solubilities of cleaned phosphor powders are negligibly low: ZnS is 68 mg per 100 ml water and Y_2O_2S is insoluble. Surface treatment of the phosphor powder is unnecessary for stabilization during storage. In addition to the stability in a PVA phosphor slurry, surface treatments will generate flicker, moiré, and smeared images.

Color phosphor screens in CRTs are produced by arranging patterned screens side by side on a faceplate. Patterned color phosphor screens are produced on the faceplate of CRTs by the application of photolithography to dried PVA phosphor screens. Phosphor screens produced by present-day screening facilities are not defect-free screens, and thus Bunsen-Roscoe's law in photochemistry is not applicable to the patterning of color phosphor screens. There are many delicate factors, which are empirically determined, that affect color phosphor screens. The reality is, however, simple. Photolithography of practical PVA phosphor screens utilizes pinholes to the faceplate in dried PVA phosphor screens [20]. Bunsen-Roscoe's law is only applicable to localized areas of pinholes. The pattern sizes and adhesion of patterned screens depend on the density and size of the pinholes in the dried PVA phosphor screens. The size and density of the pinholes are controlled by the addition of an appropriate amount of anchor particles (plate microcrystals) and small particles to the phosphor powder, as well as the drying speed of wet PVA phosphor screens. The addition of various surfactants to the PVA phosphor slurry helps by increasing the reproducibility of the generation, uniformity, and density of the pinholes. When phosphor particles have clean surfaces — both chemically and physically — we then have defect-free phosphor screens for the study of the optical and electrical properties of phosphor screens.

Only particles arranged at the top layer of the phosphor screen are involved in the generation of CL; the particles between the top layer and the faceplate are insulators. Electrons cannot flow between phosphor particles (insulators). There are electrons in vacuum, outside the phosphor particles [21]. Irradiated phosphor particles inevitably emit secondary electrons to vacuum, leaving holes in the particles. The number of holes in the particles corresponds to that of emitted secondary electrons, and secondary electrons in vacuum stay in front of the particles due to the binding force of the holes in the particles — that is, surface-bound electrons (SBEs) [22]. SBEs are sometimes called an electron cloud. The insulator does not consume the holes in the particle, such that a large number of SBEs instantly forms in front of the insulator as the incident electrons penetrate the particle. Phosphor particles are a particular insulator, which has recombination centers of pairs of electrons and holes (EHs). In phosphor particles, the holes in the particles are consumed by the recombination of EHs at the luminescence centers, thus reducing the SBEs to the number of the incident electrons. The anode collects free secondary electrons in vacuum, corresponding to the number of incident electrons [23] for closing the electrical circuit at the phosphor screen. Here there is a flicker condition of the images. If the anode field over the particles arranged at the top layer of the screen is greater than the negative field of SBEs, the anode field can conceal the negative field of the SBEs of phosphor particles [22]. The incoming electron beam reaches the phosphor particles without disturbing the trajectory of the SBEs field. If the anode fields over the SBEs are smaller than the negative field, the trajectory of the incoming electron beam is disturbed by the SBEs, generating the flicker and moiré images. The anode field at the phosphor particles, which are arranged at the top layer, changes with screen thickness, and the negative field of SBEs changes with the contamination of the surface of phosphor particles by the insulators.

Smear, low-contrast ratios, and the whitening of color images on phosphor screens are caused by the spreading of emitted CL light in a phosphor pixel to neighboring phosphor pixels by scattering [24]. Crystals of practical phosphor particles lack a center of inversion symmetry, and have a high dielectric constant ε that relates to the index of refraction n; $\varepsilon^2 = n$. Practical phosphor particles have a high index of refraction ($n = 2.2$), which is comparable to that of diamond ($n = 2.4$). CL light generated in phosphor particles get out of the particles after experiencing multireflection at the crystal boundary, and the CL light in the screen reaches the image viewer after scattering on the surface of the phosphor particles in the screen. The scattered CL light has the advantage of a wide viewing angle of light images, but also the disadvantage of spreading the CL light to neighboring phosphor pixels. If the color phosphor particles have color pigment (insulators) on the surface, the pigment may absorb some amount of different CL color light from neighboring phosphor pixels. The pigment microclusters have a limitation of absorption, and pigments do not properly work on high-light images. Furthermore, there is no pigment (and color filter) for green CL. If each phosphor pixel is surrounded by the black barrier that absorbs CL light, each phosphor pixel then confines the emitted CL light in it. Consequently, emitted CL light in the phosphor pixel does not spread to neighboring phosphor pixels in different colors [24]. A similar story applies to LCD panels.

In PL applications of phosphor powders, the phosphor powders are screened on a glass substrate on a plate and/or on the inner wall of a glass tube. The same considerations are taken into account for the screening of phosphor powders. For illumination purposes, the luminance from phosphor screens is an important concern.

Now we can design screenable phosphor powders; these powders do not contain any residual by-products. The established production techniques of phosphor powders for more than 50 years cannot produce the designed phosphor powders. The production technology of phosphor powders must be revised with the new knowledge gained in phosphor production [19, 25, 26]. These technological aspects include (1) the assignment of flux material, (2) the growth mechanism of phosphor particles by the flux, (3) the appropriate amount of flux in the heating crucible, (4) the assignment of contamination sources of oxygen, (5) the eradication of oxygen in the blend mixture charged in the crucible, (6) control of the charge density of the blend mixture in the crucible, (7) proper crucible size and shape, (8) heat programs in the furnace, and (9) posttreatment of the heated products.

This book describes various details of the above items. The content covers the spectrum from the formation of luminescence centers in particles to a highly optimized production of phosphor powders. It shows the number of excited luminescence centers in practical phosphor screens quantitatively, and that luminescence from solids can quantitatively be described by the solid-state physics developed in semiconductors. This volume also describes the electrical and optical properties of phosphor particles on the screen, in order to improve CL image quality on phosphor screens of CRTs, which may display images comparable with the images printed on sheets of paper and maximization light output of PL from phosphor screens.

2

Luminescent Properties Generated in Phosphor Particles

2.1 Excitation Mechanisms of Luminescence Centers

As illustrated in Figure 2.1, a phosphor screen is a transducer from the energy of invisible electrons (and photons) to light of visible wavelength. A good understanding of the generation mechanisms of cathodoluminescence (CL) and photoluminescence (PL) in phosphor screens may help in the optimization of luminescence properties, both theoretically and practically.

Under irradiation of ultraviolet (UV) light, phosphors emit a brilliant PL. The luminescence color is the same with CL, showing that both CL and PL are generated at the same luminescent centers in crystals. However, the excitation mechanisms of the luminescence centers differ significantly between PL and CL. We must distinguish the difference between CL and PL in the practical use of phosphors.

The luminescent color of phosphors is solely determined by the nature of the luminescence centers. A significant difference between PL and CL is their quantum efficiencies. The luminescence centers of practical PL are directly excited by photons of incident light, and the maximum quantum efficiency of PL (that is, the number of PL photons/incident photon) is 1.0. High PL intensity (i.e., a large number of PL photons) is obtained as the luminescence centers are excited with the intense excitation light, corresponding to a large absorption band of the phosphor. The energy of PL is lower than the energy of the incident photons — Stokes law or Stokes shift. On other hand, the luminescence centers of CL are predominantly excited by the recombination of pairs of electrons and holes (EHs) that are generated in the crystal by incident electrons. The quantum efficiency of practical CL (number of CL photons/incident electron) is approximately 10^3. The large quantum efficiency of CL is an advantage over PL.

A question arises as to why CL has the larger quantum efficiency. It stems from the excitation mechanisms of luminescence centers. We discuss the details of the excitation mechanisms of luminescence centers in phosphor particles under irradiation of electrons. When a crystal contains a transition

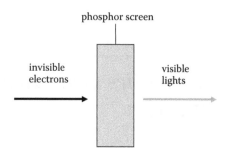

FIGURE 2.1

Phosphor screen acts as an energy transducer from invisible electrons to light of visible wavelength.

element in trace amounts, some crystals emit CL and others do not emit CL. Phosphor scientists previously believe that there was an optimal combination of luminescence center and host crystal [27, 28]. Early works to find a good combination date back to the late 1800s in France and Germany. These were empirical studies. Kroger [6] summarized these empirical works in his book in 1948. The summary does not, however, give a grip of the best combinations.

In 1968, Royce [29] and Yocom [30] invented the synthesis of Y_2O_2S:Eu phosphor powder with an application of the alkaline fusion technique of rock analysis. The Y_2O_2S crystal does not exist as a natural mineral. Their work opened the door to the study of combining luminescence centers and host crystals, using Y_2O_2S activated with many rare earths (REs) [31]. The study of Y_2O_2S:RE phosphors reveals the combinations of luminescence centers and host crystals, as well as the excitation mechanisms of luminescence centers for CL.

Under electron-beam irradiation of CL phosphors, luminescence centers in phosphors are excited in two ways: (1) direct excitation and (2) indirect excitation. The volume, at which the luminescence centers are excited, is commonly overlooked in luminescence studies on powdered phosphor screens. The penetration depth of incident electrons (~25 keV) into the phosphor particle is less than 1 μm, smaller than the particle size φ (~4 μm), so that CL generation is limited in irradiated particles arranged at the top layer of the phosphor screen. We can take an average particle (φ) at top layer for discussion of CL excitation mechanisms. Penetrated electrons in the particle scatter in the crystal via the collision of lattice ions, both elastically and inelastically. The scattering volume V_s of incident electrons in the particle (φ^2 multiplied by the penetration depth) is smaller than the particle volume V_φ ($V_s \ll V_\varphi$) and V_s is a constant with a given irradiation condition of electrons. Therefore, the number of RE ions N_{RE} in V_s is proportional to the molar fraction C ($N_{RE} = k_1 V_s C$), where k_1 is a constant. When RE ions in V_s are directly excited by incident electrons, $k_1 V_s$ is constant, and the number of excited REs

(N^*) is proportional to C. CL intensities increase with C and then decrease through a maximum by involvement of concentration self-quenching.

In the case of indirect excitation of luminescence centers by incident electrons, the incident electrons generate EHs by collision with lattice ions in V_s. Generated EHs do not directly recombine at the lattices, and they are mobile carriers in the particles [5]. Mobile EHs move out from V_s to V_φ and recombine at RE centers in V_φ. The effective volume V_{eff} that contains exciting means differs from V_s. The RE centers in V_s are also excited directly by incident electrons; however, the number of directly excited RE centers in V_s is negligibly small, $\sim 10^{-3}$ of EHs. The recombination of EHs dominates in CL intensities. If the migration of EHs is of random direction in V_φ, the number of recombinations of EHs (corresponding to CL intensities) should be proportional to φ^3 (details provided in Section 2.4). This does not occur in CL phosphors. The CL intensities linearly increase with C. This leads to the following: the migration length of mobile EHs (L_{EH}) increases with decreasing RE concentrations, and reaches particle size φ at C^*, where $L_{EH} = \varphi$. At concentrations below C^*, L_{EH} is limited by φ and L_{EH} is a constant. Therefore, V_{eff} is constant as $L_{EH} = \varphi$, but V_{eff} is variable in the range $L_{EH} < \varphi$. If CL intensities are measured as a function of C (i.e., concentration dependence (CD) curve), the curve should have two lines with different slopes (variable and constant V_{eff}), and the curve inflects at C^*. The left side of Figure 2.2 illustrates models of two different excitations of luminescence centers:

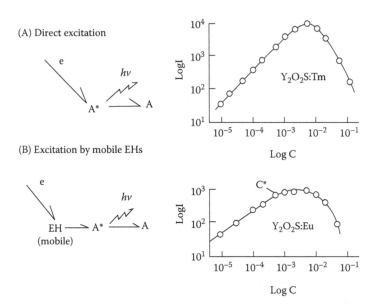

FIGURE 2.2
Difference in concentration dependence curves of CL with types of excitation mechanism of luminescence centers: (A) curve for directly excited luminescent center, and (B) curve for CL arising from the recombination of mobile EHs.

TABLE 2.1

Results of Excitation Mechanisms of RE^{3+} in Y_2O_3, YVO_4, and Y_2O_2S Crystals under Electron-Beam Irradiation, as Determined from CD Curves

RE^{3+}	Y_2O_3	YVO_4	Y_2OS_2	Remarks
		Crystals		
Eu, Dy, and Sm	RC	RC	RC	RE^{3+} and RE^{2+}
Tb and Pr	Direct	Direct	RC	RE^{3+} and RE^{4+}
Ho, Er, and Tm	Direct	Direct	Direct	RE^{3+}

Note: RC denotes recombination centers of EHs.

(A) is direct excitation, and (B) is excitation by mobile EHs [5]. Curves on the right side of Figure 2.2 show actual CD curves of CL from the Y_2O_2S:Tm and Y_2O_2S:Eu phosphors. The CD curve of Y_2O_2S:Tm is a straight line at C < 10^{-2} molar fraction, indicating that the Tm^{3+} ions in Y_2O_2S particles are directly excited by incident electrons. The CD curve of Y_2O_2S:Eu consists of two lines that inflect at $C^* = 5 \times 10^{-4}$ mole, showing that the Eu^{3+} in Y_2O_2S particles is excited by mobile EHs. Figure 2.2 shows a distinguishable difference between the excitation mechanisms of CL phosphors; that is, direct and indirect excitation. We can use the CD curves as a new tool in the study of the excitation of luminescence centers in CL phosphors.

The study is made using the REs in Y_2O_3, YVO_4, and Y_2O_2S particles. Table 2.1 provides a summary of the results. Brilliant CL is obtained with REs that are multivalent, such as Eu, Dy, and Sm (RE^{3+} and RE^{2+}), and Tb and Pr (RE^{3+} and RE^{4+}). If the RE does not have multivalences, it is directly excited by incident electrons, and results in a dim CL phosphor.

Many excitation means, involving free carriers (electrons) [12], plasmons [32, 33], and excitons [9], have been proposed for CL. Several attempts to explain the CL mechanism have been made [9, 11, 35–37]; however, solid confirmation of them has not yet been achieved. As shown in Table 2.1, the excitation means for CL are identified as EHs by the study of CD curves of CL intensity [31]. Verification of the excitation by EHs will be made by the CL photons emitted from practical CL phosphors (see Section 2.8).

The results in Table 2.1 also provide information regarding the excitation process at RE^{3+}. An RE^{3+} that has multivalences captures a moving electron (or hole), and changes the valence to RE^{2+} (via the capture of an electron) or RE^{4+} (via the capture of a hole). The RE does not release the captured carrier; it attracts a carrier of opposite charge. Subsequently, the EH recombines at the RE that first captured (trap) the carrier, releasing luminescence that is a radiative recombination of EHs. The migration distance of the EHs is determined by the distance of the first captured carrier at the RE. This is an important conclusion that determines an appropriate particle size for CL phosphors. Details of the appropriate particle size are discussed in Section 2.5.

The results in Table 2.1 provide additional information. The REs having valences RE^{3+} and RE^{4+} (Tb and Pr) are only directly excited by incident electrons in Y_2O_3 and YVO_4. The sulfur ion rather than the crystal structure is probably responsible for the indirect excitation of luminescence centers in phosphors. This is confirmed with Tb in La_2O_3 and La_2O_2S, both having the same crystal structure — hexagonal with D^3_{3d} space group. Tb^{3+} is directly excited in La_2O_3 and is excited by EHs in La_2O_2S. The results may explain why Cu (and Ag) ions, which show two valences (+1 and +2), form recombination centers for EHs in ZnS but do not form recombination centers for EHs in ZnO.

It is known that ZnS phosphors form donor-acceptor (or activator–co-activator) pair luminescence centers [10, 38, 39]. In ZnS crystals, the donor is Cl or Al and the acceptor is Cu or Ag. They have two valences (Cu^+ and Cu^{2+}, Ag^+ and Ag^{2+}, and Cl^- and Cl^{2-}, and Al^{3+} and Al^{2+}). In ZnS:Cu:Al phosphors, for example, Cu^- captures a hole generated in the valence band (S^{2-}, $3s^2$ $3p^6$) and becomes Cu^{2-}. The Cu^{2-} does not release the captured hole to the valency band (i.e., the trapped hole). On the other hand, Al^{3+} captures an electron from the conduction band (Zn^{2+}, $3d^2$ $4s^0$), becoming Al^{2+}. Al^{2+} releases the captured electron to the conduction band thermally. The electrons released into the conduction band do not directly recombine with the holes in Cu^{2+}. For recombination with the holes in Cu^{2+}, the electrons in the conduction band must trap in Al^{3+}, which distribute within an appropriate distance from Cu^{2+}. The electronic transition only occurs between Al^{2+} and Cu^{2+}, which distributes within the appropriate distance [10]. The migration distance of EHs in ZnS is determined by the average distance between Cu^{1+} ions that capture holes.

The question remains as to REs that have multivalences and do not form an efficient luminescence center in ZnS. Donor-acceptor pairs do not form in the Y_2O_2S crystal. A clarification of this point remains for future study.

2.2 Transition Probability of Luminescent Centers

As described, the luminescence centers in CL phosphors are excited by the recombination of EHs generated in the particles. The next concern then is the transition probability that an excited luminescence center will emit a photon at visible spectral wavelengths. The details of this transition probability of luminescence centers can be studied by photoluminescence (PL).

Simple quantum mechanics indicates that for luminescence centers in CL phosphors, the electrostatic dipole transition is highly possible, in which electromagnetic waves induce the alteration of the electrostatic dipole moment of an atom. The electromagnetic waves can also induce alterations of the electrostatic quadrupole and magnetic dipole moments. The transition probabilities of electrostatic quadrupole and magnetic dipole radiation are very small ($\sim 10^{-6}$) compared with that of electrostatic dipole radiation. The electrostatic dipole transition is strictly forbidden in symmetrical fields, such as in the gas and liquid phases; and the electrostatic quadrupole or magnetic dipole transition constitutes major radiation in particles that have center of inversion symmetry.

According to Laporte's rule, electrostatic dipole transitions are allowed only between electronic states of different parity [4]. For example, the ground state $^7F_2$ of odd parity in Eu^{3+} is obtained by the emission of a photon from even-parity excited states $^5D_0$ and $^5D_1$, and vice versa. The parity selection rule is lifted when the luminescence center is located at the lattice sites that lack center of inversion symmetry. The probability of allowed electrostatic-dipole transitions in an asymmetric field is significantly greater than that in a symmetric field. In some cases, the ground state has a spin-singlet character, while the lowest excited (emitting) state has a spin-triplet character. The spin-forbiddenness of electronic transitions between these two states is relaxed through spin-orbit interaction with the high-lying singlet excited states. The spin-orbit interaction becomes important in heavy elements such as REs.

Crystals of commercial phosphors, such as ZnS, (Zn, Cd)S, ZnO, Y_2O_2S, Y_2O_3, and others, do not have inversion symmetry. In practical phosphor crystals, the multivalence element, which forms luminescence centers, occupies cationic lattice sites that lacks center inversion symmetry. A combination of host crystal, which lacks a center of inversion symmetry, and a multivalence element allows electrostatic dipole transitions in practical CL and PL phosphors. The radiative transition probability of the excited luminescence centers varies with the combination. This variation in transition probability can be studied by PL under direct excitation of the luminescence centers. PL intensities from phosphor screens are usually measured under UV light, which is directly absorbed by the luminescence centers.

A commonly overlooked concern in the study of PL is the volume (V_{eff}) into which the incident light penetrates the layers of the phosphor screen. Phosphor screens are constructed of layers of phosphor particles. Crystals that lack center of inversion symmetry have a high dielectric constant ε. The value ε relates to the index of refraction n; $\varepsilon^2 = n$. Table 2.2 gives the index of refraction of host crystals of typical CL phosphors, together with that of diamond as a reference. The n values of ZnS ($n = 2.38$) and Y_2O_2S ($n = 2.16$) are comparable with that of diamond ($n = 2.41$).

TABLE 2.2

Index of Refraction of Typical Commercial Host Crystals and Diamond

Host Crystals	Index of Refraction, n
ZnS	2.38
Y_2O_2S	2.16
Zn_2SiO_4	1.72
Diamond	2.41

In powdered phosphor screens, therefore, a large fraction of incident light (~40%) is reflected from the surface of individual particles, with the remainder penetrating the phosphor surface to generate PL. The reflected UV photons entering the screen are randomly scattered throughout the particle structure. Penetration of incident UV light into a deep layer in the screen is achieved using scattered UV light on the surface of the particles, although phosphor particles have a strong absorption coefficient.

We can calculate the UV light absorbed by the phosphor screen in various layers of phosphor particles [40]. When light of intensity I_o irradiates the phosphor screen constructed of L layers of particles, the absorbed UV light on the screen (I_{abs}) is given by subtracting the light intensity transmitted from the screen (I_t): that is, $(I_o - I_t)$. I_t is given by

$$I_t = I_o \, e^{-(\alpha + \beta)L} \tag{2.1}$$

where α and β bare the absorption and scattering coefficients per phosphor layer, respectively, and they are different from the absorption and scattering values determined from single crystals. Thus, we have:

$$I_{abs} = k_2 \, I_o\{1 - e^{-(\alpha + \beta)L}\} \tag{2.2}$$

where k_2 is constant. Assuming that a UV photon absorbed by the phosphor particle generates a PL photon, PL intensities measured on phosphor screens in the reflection mode (irradiation side) are given by Equation (2.2).

If PL intensities are measured in terms of transmitted light from the phosphor screen (I_{trans}), the emitted light at a deep layer in the screen loses its intensity with horizontal scattering in the phosphor screen. I_{trans} is given by (1) the light intensity emitted in the given phosphor layer and (2) the scattering loss of the emitted light by the layers between the emitted and output layers. I_{trans} is expressed as

$$I_{trans} = k_2 \, I_o\{1 - e^{-(\alpha + \beta)L}\} - k \, I_o\{1 - e^{-\beta}\}$$

$$\times \, [\{1 - e^{-(L-1)(\alpha + \beta)L}\} \, e^{-(L-1)\beta} + \{1 - e^{-(L-2)(\alpha + \beta)L}\} \times e^{-(L-L+1)\beta}]$$

$$\times \ldots \tag{2.3}$$

The α and β values of phosphor screens are experimentally determined with commercial red Y_2O_3:Eu phosphor powder under 254-nm UV irradiation. The values are $\alpha = 0.26$ and $\beta = 0.28$ per phosphor layer. Figure 2.3 shows the PL light intensity of the Y_2O_3:Eu phosphor screen with transmission and reflection modes calculated from Equations (2.2) and (2.3) (solid lines) and experimental data (dots). If the PL intensities are measured in the reflection mode (I_{ref}), a reliable PL I_{ref} is obtained with the screen thicker than ten layers. For a determination of QE, the measurements of the PL of practical phosphors should be made with I_{ref} on a phosphor screen thicker than ten layers.

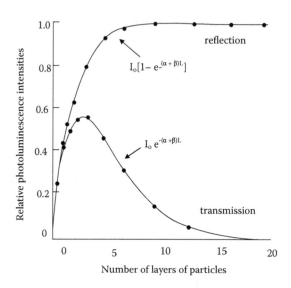

FIGURE 2.3
Theoretical (solid lines) and experimental (dotted) curve of PL intensities from phosphor screens in reflection and transmission modes, as a function of the number of layers of phosphor particles.

2.3   Short Range, Rather than Long Range, Perfection for Phosphors

It is believed in the phosphor community that the radiative transition probability of excited luminescent center is influenced by the crystal perfection of phosphor particles. Microcrystals have high crystal perfection, and polycrystals exhibit poor perfection with many crystal boundaries. Accordingly, microcrystals can have a high transition probability. Crystal perfection around luminescence centers, which influences the transition probability, can be studied in the ZnS:Cu:Al phosphor.

A band model consisting of a valency band and a conduction band is usually used in discussing the luminescent properties of the ZnS:Cu:Al phosphor [10]. Figure 2.4 shows the band model that explains the luminescence process from (1) excitation of electron from the valency band, (2) migration of the excited electron in the conduction band, (3) the trapped electron in the donor level (Al^{2+}), (4) migration of the hole in the valency band, (5) the trapped hole in the acceptor level (Cu^{2+}), and (6) the electron transition (radiative) from the donor to the acceptor. The green luminescence of the ZnS:Cu:Al phosphor is well explained by the band model as far as the peak wavelength of the luminescence band is concerned.

Here arises a question. According to the band model, the green luminescence should be a line spectrum; however, the green luminescence of the

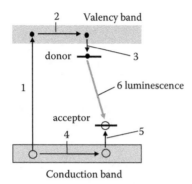

FIGURE 2.4
Explanation of electron and hole transportation (1–5) and radiative recombination (6) between donor and acceptor levels of the ZnS:Cu:Al green phosphor.

ZnS:Cu:Al phosphor is a broad band of Gaussian shape (half-width = 0.33 eV). The band model cannot explain the broad band. The band model provides important information such that radiative recombination occurs between electrons trapped in the Al^{2+} (donor) and holes trapped in the Cu^{2+} (acceptor). Figure 2.5 illustrates an atomic model of the radiative transition in the ZnS crystal. The separation distance (r) of Al^{2+} and Cu^{2+} determines the energy of the emitted photon E(r), which is given by [41]

$$E(r) = Eg - (E_D + E_A) + e^2/\varepsilon_o\, r \qquad (2.4)$$

where E_D is the depth of the *D* level below the conduction band (0.03 eV) and E_A is the depth of the *A* level above the valency band (1.2 eV), e is the electron charge, and ε_o is the dielectric constant of ZnS in vacuum.

Given the deep level of the acceptor (1.2 eV), the trapped holes in the acceptor do not release to the valency band. The transition distances (energy) of the *D-A* pairs are determined from the acceptor. The luminescence centers are formed by random combinations of *D-A* pairs in the ZnS crystal, and the

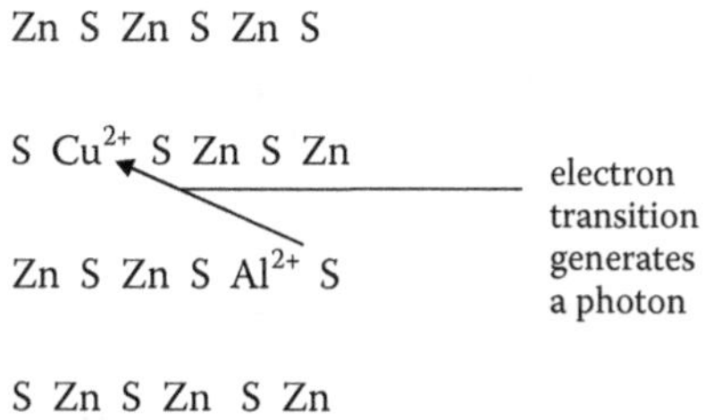

Zn S Zn S Zn S

S Cu^{2+} S Zn S Zn

Zn S Zn S Al^{2+} S

S Zn S Zn S Zn

electron transition generates a photon

FIGURE 2.5
Atomic model of radiative electron transition from Al^{2+} to Cu^{2+} in ZnS lattices.

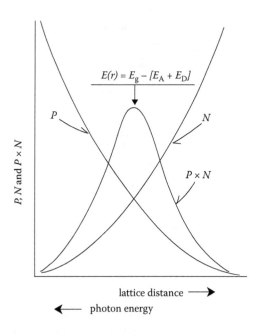

FIGURE 2.6
The number of available lattice sites (N), the transition probability (P) between the *D-A* pair recombinations, and the luminescence spectrum as the result of the product of (N × P).

possible combinations are proportional to N^3, where N is the number of cation lattice sites from the A ion (i.e., the acceptor that has captured a hole). The transition probability (that the *D-A* pair is emissive) decreases as the separation distance increases [38, 39]. Figure 2.6 quantitatively illustrates the number (N) of available donors in the lattice sites, the transition probability (P), and the luminescence spectrum as products of (P × N), which is bell shaped (i.e., Gaussian). The measured luminescence spectrum of the ZnS:Cu:Al phosphor is certainly Gaussian, with 0.33 eV half bandwidth. Because the luminescence band is caused by the distribution of separation distances of *D-A* pair recombination centers, the luminescence band is less dependent on phosphor temperature. The decay time is an inverse function of transition probabilities (P^{-1}). As the excitation intensities increase, de-excited *D-A* pairs are involved in the luminescence process, resulting in an enhancement of photons at short wavelengths in the luminescence spectrum. This gives rise to a shift in the luminescence spectrum to shorter wavelengths. The spectrum of the afterglow also shifts to longer wavelength after termination of the excitation [38, 39, 42]. The luminescence band of ZnS:Cu:Al phosphors is fully explained by an atomic model. The spectrum supplies other important information as well — for example, the maximum interaction distance of *D-A* pairs.

The maximum interaction distance of *A-D* can be estimated from the maximum width of the spectrum of the ZnS:Cu:Al phosphor. It is 0.8 eV,

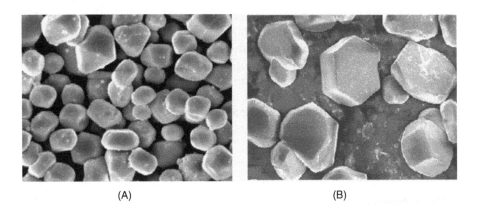

(A)   (B)

FIGURE 2.7
SEM images (2000X) of polycrystals (A) and microcrystals (B) of Y_2O_2S:Eu red phosphor particles
that give the same CL intensities.

corresponding to the maximum separation distance of 500 Å. The lattice
constant of cub-ZnS is $a = 5.4$ Å; 500 Å corresponds to about 100 lattice sites.
Therefore, the A ion may have a chance to interact with a D ion that is
distributed in a spherical volume of 500 Å (= 50 nm) diameter. The required
crystal perfection for a D-A pair recombination center is within the sphere
of the 50-nm radius, not the long-range perfection in the particles. The
electronic states of luminescent centers are governed by the localized crystal
perfection at around the center. Phosphor particles in nanometer sizes may
have a high transition probability. It follows then that CL properties are not
influenced by the particle sizes and shapes of practical phosphor powders.
Since phosphor powders are separately produced as polycrystals and microcrystals (Figure 2.7), both phosphor powders emit the same CL intensity and
color. Accordingly, the produced phosphor powders, which contain different
shapes and sizes of particles, give the same CL intensities. This ensures that
a large number of CL phosphor particles (10^{17}) can be produced at once with
a simplified production process. The variation of CL intensities (±10%) is
not caused by the transition probability of luminescence centers; it comes
from the additional luminescence centers that form in phosphor particles as
contamination.

Practical PL phosphors reveal a story that is similar to ZnS luminescence
bands. In PL phosphors, consideration is given to the excitation band. The
luminescence centers in efficient PL phosphors are directly excited through
the charge-transfer band, which is formed by electron transfer from an activator to neighboring anions. The charge-transfer band is given by the products of the number of available anions, multiplied by the transition
probability. This gives a broad band, similar to the luminescent bands of ZnS
phosphors. The magnitude of the absorption coefficient of the charge-transfer
band corresponds to the absorption coefficient of the host lattice. By studying single crystals, the absorption coefficient D (log absorption distance

cm^{-1}) is about 5 in many host crystals, corresponding to a penetration distance of 0.1 μm from the surface of phosphor particles. The generation of PL
in practical phosphors is restricted in the surface volume within 0.1 μm depth
from the surface of individual phosphor particles in the screen. The PL
intensities of PL phosphor particles are not influenced by the size and shape
of phosphor particles. The variation in PL intensity of practical phosphor
screens derives from surface contamination, which absorbs the exciting light
of PL and the absorption of incident light of individual particles in the
phosphor screens.

2.4   Number of Luminescence Centers Involved in the Measurement of Luminescence Intensities

In practice, the number of luminescence centers (N_{Lum}) involved in the generation of luminescence is an important concern for brighter phosphor powders. Total N_{Lum} in phosphor particles is proportional to the cube of the
particle size $\varphi(N_{Lum} = k\varphi^3)$. The measured CL and PL intensities of the phosphor screen (I_{Lum}), as a function of particle size, do not change with φ^3,
showing that I_{Lum} is not directly related to N_{Lum} in V_{φ}. In general, the luminescence intensities linearly increase with increasing concentration C (molar
fraction) of the luminescence centers in a low concentration range, and then
decrease due to concentration quenching. The molar fraction C is the ratio
of N_{Lum} to the number of cation lattice sites in a finite volume (constant). No
one questions the linear dependence of I_{Lum} from phosphor screens in low
C regions, which is analogous to absorption measurements in chemical analysis, where the Beer-Lambert law ($e^{-\alpha C}$) is applicable to absorption α is the
absorption coefficient of the solution). The Beer-Lambert law is only applicable to absorption in the cell for which the pathlength (usually 1-cm long)
is shorter than the length determined by $1/\alpha$. The Beer-Lambert law is not
applicable to the absorption in a cell (e.g., 5 cm) longer than the length
determined by $1/\alpha$, of which incident light is totally absorbed in the cell;
i.e., infinite volume. Light absorption by luminescence centers corresponds
to the luminescence intensity from the phosphor particles. If an infinite
volume is involved in the excitation of luminescence centers in the phosphor
screen, all photons of the exciting light are absorbed by the luminescence
centers in the screen. The absorbed light should transduce to PL, and the
screen emits a constant I_{Lum} with different C. This is not reality. A linear
dependence of I_{Lum} on C is only obtained with measurements of PL from the
phosphor screen, which the exciting light penetrates through a finite volume
(penetrated layers multiplied by cross-section of incident lights) in the phosphor screen, and the luminescence from the finite volume of the screen is
detected by the photometer. It should note that we are here discussing the
finite volume where the incident lights have penetrated in the screen by light

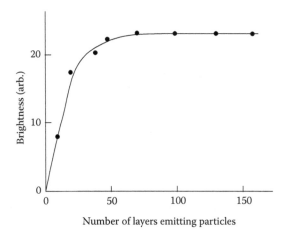

FIGURE 2.8
Luminescence intensities from screen constructed with particles, for which each particle uniformly adheres radioactive Pm[147] (beta-rays) microclusters on the surface.

scattering on the surface of phosphor particles, and the finite volume differs from the absorption volume that individual particles in the screen have absorbed the incident lights.

As described in Section 2.2, the penetration depth of incident UV light into the phosphor screen is calculated from Equation (2.1); the UV light can penetrate into a deep layer of the phosphor screen with scattering on the surface of particles in the screen, and individual phosphor particles in the layers absorb the incident light. Incident light having a small α value penetrates into layers deeper than the layers at which emitted PL is detected. If this is so, then there is a finite volume (V_{eff}) in the phosphor screen for I_{Lum} measurements of PL. The maximum finite volume can be studied in phosphor screens constructed with using ZnS:Cu:Cl phosphor particles onto which radioactive Pm[147] (beta-rays) uniformly adhere [43]. Each phosphor particle emits an equal intensity of CL. Figure 2.8 reveals the experimental results. The measured I_{Lum} from the screen increases with the number of layers, and saturates when the number of layers is greater than 70. If the phosphor screen is thicker than 70 layers of particles, a reliable PL measurement of I_{Lum} will be obtained under irradiation of light having a small α value. Figure 2.9 illustrates a model of the finite volume in a phosphor screen for I_{Lum} measurements.

The effective volume V_{eff} of a PL study on phosphor screens is given by the cross-section of the incident light beam, multiplied by the 70 layers of phosphor particles. N_{Lum} at constant V_{eff} is given by

$$N_{Lum} = k_3 C \, V_{eff} = k_4 C \tag{2.5}$$

where k_3 and k_4 ($= k_3 \, V_{eff}$) are constant. Luminescence centers occupy lattice sites and they are fixed in position. The centers are sampled by moving (scattered) light in the screen. According to probability theory, the number

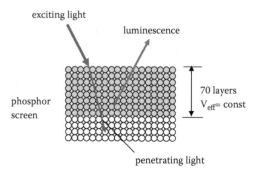

FIGURE 2.9
Model for the measurement of reliable CD curves of PL on phosphor screens.

of excited luminescence centers N^* in V_{eff} is a consequence of the random sampling of N_{Lum} in V_{eff} by the exciting light J. Sampling results in

$$N^* = N_{Lum}! \; J! \; /(N_{Lum} - 1)! \; (J - 1)! = J \; N_{Lum} = k_5 CJ \qquad (2.6)$$

Under a given exciting light (J = constant),

$$N^* = k_6 C \qquad (2.7)$$

We can use C as the concentration of luminescence centers in V_{eff}.

To figure out a complete PL study from phosphor screens, one should consider concentration quenching mechanisms, which are involved in a high C region.

The concentration quenching mechanisms of luminescence from the $^5D_0$ emitting level of Eu^{3+} are well studied [5]. They are electrostatic Q-Q, Q-D, and D-D interactions, and magnetic D-D interactions, where Q and D represent quadrupole and dipole, respectively. Electrostatic interactions occur by overlap of the wave functions of the emitting state $^5D_0$ of Eu^{3+} ions. They occur in the low concentration region (C < 10^{-3} molar fraction), in the order of Q-Q < Q-D < D-D. Magnetic D-D interactions occur by the overlapping of spin orbits of the ions that occupy the nearest-neighbor cation sites (C > 1×10^{-3}). The number of isolated (emissive) Eu^{3+} ions (that do not have magnetic D-D interaction) in a particle is calculated from the crystal structure. The luminescence line at 611 nm (transition from $^5D_0$ to $^7F_2$ of Eu^{3+}) has magnetic D-D interaction. The fraction (F) of isolated Eu^{3+} in the crystal is given by

$$F = (1 - C)^z \qquad (2.8)$$

where z is the weighted number of the nearest-neighbor cation sites. Because $N^* = k_6 C$, I_{Lum} from emissive $^5D_0$ Eu^{3+} in the phosphor is expressed as

$$I_{Lum} = FN^* = k_6 C \; (1 - C)^z \qquad (2.9)$$

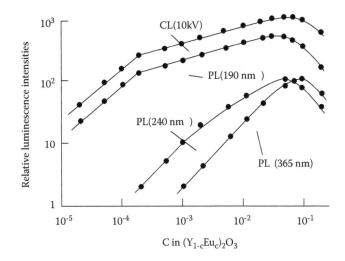

FIGURE 2.10
CD curves of $^5D_0$ Eu^{3+} emission from Y_2O_3:Eu red phosphors as a function of Eu concentration in the range from $C = 1 \times 10^{-5}$ to 2×10^{-1} mole fraction. Phosphor screens (1-mm thick) are irradiated with the UV light of 365, 240, and 190 nm and with 10-kV electron beam. Relative intensities are only applicable for each CD curve.

Equation (2.9) represents I_{Lum} as a function of concentration dependence (CD) over the entire C range in V_{eff}. When C is less than 10^{-3}, the I_{Lum} of the PL is proportional to C. The optimum concentration C_{opt} for $^5D_0$ Eu^{3+} emission from the phosphor is given by differentiating Equation (2.9). Because $z = 9$ for Y_2O_3, the C_{opt} of the Y_2O_3:Eu phosphor is $(1 + 9)^{-1} = 0.10$ mole fraction. Figure 2.10 shows the CD curve of $^5D_0$ Eu^{3+} emission from Y_2O_3:Eu red phosphors under irradiation of 365-, 240-, and 190-nm UV light and an electron beam of 10 keV, in the range from $C = 1 \times 10^{-5}$ to 2×10^{-1}. The relative intensities are only applicable to each CD curve. CD curves and maximum C in Figure 2.10 markedly differ with UV light, which have different α values. The α–values of the given exciting light increase with C. The relative magnitudes of the α–values of UV light of the phosphors can be estimated from the excitation spectrum of the emission from $^5D_0$ Eu^{3+} (611 nm), which is measured on powdered phosphor screens.

As shown in Figure 2.11, the Y_2O_3:Eu (0.001 molar fraction) phosphor has a negligibly small α-value with the 365-nm light. The CD curve of PL (365 nm) fits on the curve calculated from Equation (2.9), indicating that the PL measurements are made with constant V_{eff}. If the phosphor screen is excited with 240-nm light, the penetration depth of the UV light at high C (>1 × 10⁻⁴ molar fraction) becomes less than 70 layers, and the V_{eff} decreases with C. Because V_{eff} decreases with C, the CD curve in the high C range does not fit Equation (2.9), and the maximum Eu concentration apparently shifts to a lower concentration. The results in Figure 2.10 indicate that the quantitative study of PL on powdered phosphor screens must be conducted under

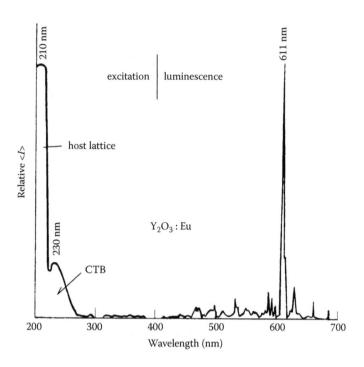

FIGURE 2.11

Emission and excitation spectra of Y_2O_3:Eu (0.001 molar fraction) red phosphor. Excitation spectrum is obtained with the emission line at 611 nm.

irradiation of light with a small α value, and the screen should be thicker than 0.3 mm ($= 4 \times 10^{-4}$ cm $\times$ 70 layers).

Light of 190 nm is absorbed by crystal lattices, and the absorption coefficient of crystal lattices is usually $D_{ab} = 10^5$ cm^{-1}, corresponding to a penetration depth of 0.1 μm (1 cm/10^5), shallower than particle size $\varphi = 4$ μm). The 190-nm light is absorbed by the particles arranged at the top layer that is exposed to the incident light. The light does not penetrate into particles lying beneath the exposed particles. The absorbed light generates EH pairs in the exposed particle, and the EHs recombine at luminescence centers in the particles. The same luminescence mechanism of CL involved in PL by 190-nm light on phosphor screen as seen in Figure 2.10.

The results of CD curves in Figure 2.10 clearly show the difference in the luminescence process between CL and PL in the same phosphor particle, although the luminescence process occurs at the same luminescence center in the phosphor particle. We now discuss the V_{eff} of CL {and PL (190 nm)} for a particle in the phosphor screen.

As described in Section 2.1, the CL in practical phosphors is generated by the radiative recombination of EHs. The EHs in CL phosphor particles migrate in quasi-one direction along the electric field that is internally generated photopiezoelectrically [5]. When the EH meets one luminescence center on the migrating path, the luminescence center is excited by EH for CL generation. The migration distance of EHs (L_{EH}) is an inverse function of C; L_{EH} increases as C decreases. L_{EH} reaches particle size at C* where $L_{EH} = \varphi$. In the range that L_{EH} is greater than φ, $V_{eff} = V_\varphi$ = constant. No concentration quenching mechanism is involved in this C region. N* in $V_{eff} = V_\varphi$ is proportional to C. Therefore, L_{Lum} is given by

$$L_{Lum} = k_7 C \quad (L_{EH} > \varphi) \tag{2.10}$$

In the range that L_{EH} is less than φ, V_{eff} shrinks with C from V_φ; $V_{eff} < V_\varphi$. The number of cations along the quasi-one directional path in V_φ is given by $(V_\varphi)^{1/3}$. Therefore, C in the range that L_{EH} is less than φ is given by $(V_{eff})^{1/3}$. Then, N* in V_{eff} ($< V_\varphi$) is given by

$$N^* = k_8 \, (V_{eff})^{1/3} = k_9 C^{1/3} \tag{2.11}$$

$$L_{Lum} = N^* \, F = k_9 C^{1/3}(1 - C)^z \quad (L_{EH} < \varphi) \tag{2.12}$$

Figure 2.12 shows CD curves of the CL of $^5D_0$ Eu^{3+} from YVO$_4$:Eu (z = 4), Y$_2$O$_3$:Eu (z = 9), and Y$_2$O$_2$S:Eu (z = 12) phosphors in a practical concentration

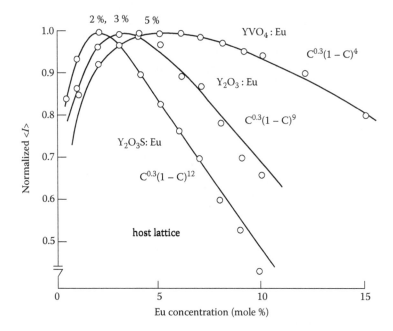

FIGURE 2.12
CD curves of CL from YVO$_4$:Eu (z = 4), Y$_2$O$_3$:Eu (z = 9), and Y$_2$O$_2$S:Eu (z = 12) phosphors in practical concentrations between 0.5 and 15 mol%. CL data are measured with 15 kV, 1 μA cm^2. Each curve is normalized at peak intensity.

region between 0.5 and 15 moles. All CD curves in Figure 2.12 fit the calculated curves from Equation (2.12) with different z values, showing that CL is generated by the radiative recombination of EHs at Eu^{3+}. The results in Figure 2.12 indicate that the optimal Eu concentrations of CL from Y-compound phosphors are calculable using Equation (2.12).

2.5 Optimal Particle Sizes for Practical Phosphors

In designing CL phosphors, we have shown that crystal perfection of phosphor particles is not required for phosphor production. Thus, optimal particle size φ_{opt} becomes an important concern in the production of phosphor powders. A clue as to the determination of φ_{opt} can be found from the CD curve of CL in Figure 2.2, that is, the C* that gives the maximum migration distance of EHs in phosphor particles. In the determination of C*, there are two difficulties associated with practical powdered phosphor screens: (1) the reproducibility of sample preparations, and (2) the measurement conditions of CD curves.

To obtain a CD curve, one should prepare at least 15 samples with different C values. The phosphor powder contains 10^{11} particles per gram. An experimental sample is at least 100 g, containing 10^{13} particles. In determining the C* of CD curves, 10^{13} particles should have equal size and shape. A high reproducibility of size and shape of the particles is required in sample preparation. Figure 2.13 shows microscopic pictures (600X) of particles of a commercial phosphor powder (A) and a suitable phosphor powder (B) for C*

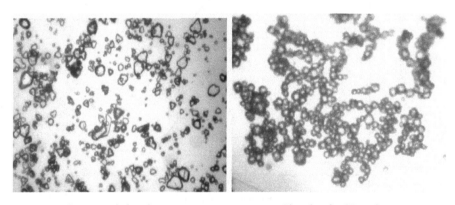

Commercial phosphor   Phosphor for CD study

FIGURE 2.13
Microscope image (600X) of particles of a commercial phosphor powder and appropriate phosphor powder for determination of C* values of CD curves.

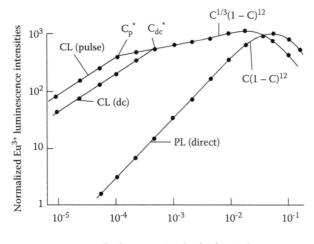

FIGURE 2.14
Normalized CD curves of Y_2O_2S:Eu phosphor. Curves of CL (pulse) and CL (dc) obtained under irradiation of pulsed (2 ns) and continuous (dc) electron beams (15 kV, $1\mu A \cdot cm^{-2}$). The CL curves inflect at C_p^* and C_{dc}^*, respectively. Curve of PL (direct), which fits to curve by Equation (2.9), is obtained under 414-nm irradiation.

studies of CD curves. The samples prepared with the established process are the same as for commercial phosphor powders, for which the particles are widely distributed in both size and shape. One can obtain the CD curve with commercial powders, with an indistinct determination of C^*. The distinct determination of C^* of CD curves is only obtained with the phosphor powder shown on the right side of Figure 2.13. The details of the preparation of the phosphor powders are described in Chapter 6. Although the sample preparations have high reproducibility, the determination of C^* is a statistical consequence of a huge number of phosphor particles, and is not from a single crystal.

The CL gives two different C^* values, depending on the irradiation conditions of electron beam on the same phosphor screen; that is, pulsed (2 μs) and continuous (dc) irradiation. Figure 2.14 shows normalized CD curves of the CL (pulse) and the CL (dc) from the same Y_2O_2S:Eu phosphor screen, together with the PL (direct) as verification of Equation (2.9). The CD curves of CL inflect at C_p^* and C_{dc}^*, respectively. The difference between C_p^* and C_{dc}^* for the same phosphor screen is explained in terms of statistics. The risetime of the CL in Eu^{3+} is faster than 0.1 μs, and the lifetime of excited Eu^{3+} is 500 μs. During a 2-μs pulse irradiation of 60 Hz (17 ms interval), the Eu^{3+} ion on the migrating path of EH is only excited once by EH, and returns to the ground state before the next irradiation time (500 μs << 17 ms). According to probability theory, this is sampling without replacement [44]. Under dc irradiation, the EHs continuously generate in the same phosphor

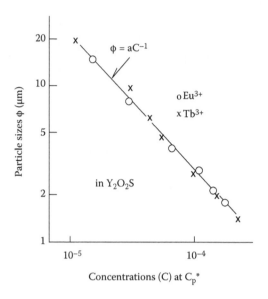

FIGURE 2.15
Average particle size φ vs. C_p^* (mole fraction) of CD curves of CL intensities of Y_2O_2S:Eu (o) and Y_2O_2S:Tb (x) phosphors.

particles. The excited Eu^{3+} on the migration path returns to the ground state. The returned Eu^{3+} has a chance to meet a recently generated EH on the path (i.e., sampling with replacement) [44]. Sampling with replacement apparently increases C on the migration path. Therefore, C_p^* is less than C_{dc}^* for the same particle. For the determination of $L_{EH} = \varphi$, we use C_p^*.

Assuming that the migration distance of EHs corresponds to $L_{EH} = \varphi$ at C_p^*, the C_p^* changes with φ. Figure 2.15 shows average particle sizes φ versus the C_p^* of CD curves of Y_2O_2S:Eu (o) and Y_2O_2S:Tb (x). Eu^{3+} and Tb^{3+}, respectively, detect the L_{EH} of electrons and holes in the Y_2O_2S particle [45]. The obtained data fit a straight line, showing a linear relationship between φ and C_p^* in Y_2O_2S particles. The L_{EH} of both the electrons and the holes is expressed by

$$L_{EH} = a/C_p^* \qquad (2.13)$$

where a is the distance between nearest cations. The value of a is approximately 3.5 Å in many phosphor crystals.

As shown in Figure 2.12, practical Y_2O_2S phosphors are produced with C_{opt} a few mol %. The calculated L_{EH} value of practical Y_2O_2S:Eu phosphors is 0.02 μm (= 20 nm), ($<<\varphi$). Therefore, practical Y_2O_2S:Eu phosphors have a constant CL intensity with different φ.

The migration distance of EHs in practical ZnS blue and green phosphors is another story. In ZnS phosphors, the maximum concentration of Cu (or Ag) in ZnS crystals is limited by the diffusion of Cu (or Ag) in the ZnS lattice

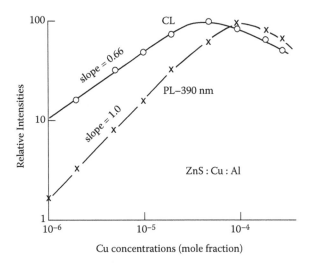

FIGURE 2.16
CD curves of CL and PL (390 nm) of ZnS:Cu:Al blue phosphor.

(C = 1 × 10⁻⁴). As Cu (or Ag) adds beyond the limit of diffusion, the excess Cu (or Ag) occupies interstitial lattice sites that are not from luminescence centers, giving rise to a gray body color. The gray coloration significantly decreases the CL intensity. Figure 2.16 shows the CD curves of CL (15 keV, 1 μA cm⁻²) and PL (under 390-nm light) of ZnS:Cu:Al phosphors. We can take $C_{opt} = 1 \times 10^{-4}$ mole as the optimal particle size of ZnS phosphors.
The calculated L_{EH} of practical ZnS phosphors is

$$L_{EH} = \varphi = a/C_{opt} = 3.5 \times 10^{-8} \text{ cm}/1 \times 10^{-4} = 3.5 \ \mu m \qquad (2.14)$$

The CL intensities of practical ZnS phosphors increase in increments of φ up to $\varphi = L_{EH} = 3.5 \ \mu m$. In this range of particle size ($\varphi < L_{EH} = 3.5 \ \mu m$), some migrating EHs in the particle do not have a chance to meet a luminescence center, and thus recombine nonradiatively at surface recombination centers, which are possibly anion vacancies. When $\varphi > L_{EH} = 3.5 \ \mu m$, all of the generated EHs in the phosphor particles meet luminescence centers in the particle, giving rise to a constant CL intensity. Therefore, the CL intensities of ZnS phosphors increase with particle size up to 3.5 μm and then saturate at $\varphi > 3.5 \ \mu m$.
Figure 2.17 shows the curve of relative $C^{1/3}$ and the measured CL intensities of ZnS:Cu:Al green and Y₂O₂S:Eu red phosphors as a function of φ between 1 and 5 μm. The curves are normalized at $\varphi = 1 \ \mu m$. CL intensities of ZnS:Cu:Al phosphor increase with φ up to 3 μm, and then approach a constant value of 1.5 at $\varphi = 4 \ \mu m$. By considering the distribution of the sizes and shapes of the particles (not spherical), $\varphi = 4 \ \mu m$ is in good agreement with $L_{EH} = 3.5 \ \mu m$ as calculated from Equation (2.14). Similar results were

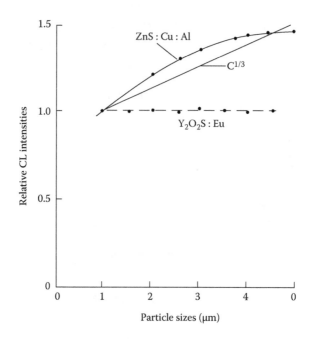

FIGURE 2.17
Relative CL intensities of ZnS:Cu:Al and Y_2O_2S:Eu phosphors and $C^{1/3}$ as a function of particle size φ. Curves are normalized at $\varphi = 1$ μm.

also obtained with ZnS:Ag:Cl blue phosphors and the Y_2O_2S:Eu (1×10^{-4} molar fraction) phosphor. Therefore, the curve of ZnS:Cu:Al phosphors in Figure 2.17 generalizes as the curve of phosphors that have a low C of luminescence centers. Practical Y_2O_2S:Eu phosphors, as anticipated, maintain constant CL intensities with φ. In screening of color CRT production, it is preferable to have the same screening conditions with different colored phosphors. Then, φ of Y_2O_2S phosphor adjusts to the size of the ZnS phosphor (i.e., 4 μm).

Note that the particle sizes for EH migration should use average φ determined under a microscope. Many commercial instruments for particle size determination measure weight, surface area, and the volume of the settling particles in suspension. The measurement results are equivalent particle sizes, and differ from φ as determined by length under a microscope. The equivalent particle sizes are larger than the sizes determined by a microscope (about 2 times φ) [3].

Also note that there have been several reports of CL from nanoparticles [46–59]; however, phosphors from nanoparticles are not applicable to practical display devices. Particle phosphors of nanometer size do not exhibit practical luminance. The nanoparticle (e.g., 30 nm size) of ZnS:Ag:Cl or ZnS:Cu:Al contains only three luminescence centers in the particle. If an electron having an energy of 500 eV irradiates onto the nanoparticle, the penetrated electron generates 130 EHs in the nanoparticle. The CL intensities of the ZnS nanoparticle are saturated by the penetrated electron. The nanophosphor particles of ZnS:Mn and Y_2O_3:Eu contain $N_{Lum} = 10^3$, allowing seven incident electrons for saturation. The electron beam in 1 nA contains 10^{10} electrons. The excited luminescence centers in the nanoparticle are saturated with incident electrons of 10^{-16} A. Although the study of nanophosphor particles may provide scientific information about luminescence centers, the CL intensities from nanoparticles lag behind in terms of practical application in CL devices.

2.6 Stability of Luminescence Centers in Phosphor Particles

Luminescence centers, which are impurities in the host crystal, should be stable in the crystal during use. As long as the crystals (1) maintain electric neutrality and (2) are deformation-free, the crystals will remain stable. If the impurities have larger and/or smaller ionic radii than the lattice ions, then the crystal has a lattice deformation due to the impurities. The unstable impurities diffuse from the crystal during cool-down from growing temperature of the particles (~1000°C), and slowly diffuse out under operating temperature. Material science determines the conditions of the stability of the diffused impurities in crystals. If the ionic radii of the impurities fall within ±10% of the radii of the host lattice ions, then the lattice deformation is small and the impurities can remain in the crystals.

Take a look at the typical luminescence centers of Cu^+ and Cl^- pairs in a ZnS phosphor. Table 2.3 gives the ionic radii of the activators, co-activators, and lattice ions of ZnS. There is a small difference between S^{2-} and Cl^- (98%). ZnS crystal has intrinsic Zn vacancies (1×10^{-4} mole fraction of Zn lattice) [3]. Each vacancy has two positive charges. For electrical neutrality (charge compensation), Cl^- smoothly diffuses into the S^{2-} site of the ZnS crystal with negligible lattice deformation. There is a large difference between Zn^{2+} and Cu^+ (130%). Cu^+ diffuses into the ZnS crystal by filling Zn vacancies up to 1×10^{-4} mole fraction [5], deforming the lattice (30%). If ZnS already contains Cl^-, the diffused Cu^+ is somewhat stabilized by the charge compensation of Cl^-.

TABLE 2.3

Ionic Radii of Typical Activators and Co-activators in ZnS Phosphors

Cations	Radii(Å)	Difference(%)	Anions	Radii(Å)	Difference(%)
Zn^{2+}	0.74	0	S^{2-}	1.84	0
Cu^{1+}	0.96	30	Cl^-	1.81	2
Ag^{1+}	1.26	70	O^{2-}	1.36	26
Al^{3+}	0.56	31			

The large lattice distortion is relaxed by incorporating O^{2-} into S^{2-} sites. O^{2-} has ionic radius (1.36 Å) that is 74% of that of S^{2-} (1.84 Å). Cu^+ is 130% of Zn^{2+} and O^{2-} is 74% of S^{2-}, so that the average lattice distortion is 102% {= (130 + 74)/2}. Hence, Cu^+ in the ZnS crystal is stabilized by incorporating both Cl^- and O^{2-}. That is, the ZnS:Cu:Cl phosphor is smoothly contaminated with O^{2-}. The contaminated O^{2-} forms an additional luminescence center associated with Cu^{2+}, which gives a luminescence band on the long-wavelength side of the Cu–Cl luminescence band. Furthermore, the Cu–O luminescence centers give a long luminescence decay time. The ZnS:Ag:Cl phosphor is also smoothly contaminated by O^{2-} for the same reason, resulting in a large lattice distortion (44%).

Perfect compensation of both electric charge and lattice distortion of ZnS:Cu green phosphor can be achieved by incorporating Al^{3+} on Zn^{2+} sites. Cu^+ is 130% and Al^{3+} is 69% of Zn^{2+} (0.74 Å), so that the average lattice distortion is (130 + 69)/2 = 100%. Hence, Cu^+ and Al^{3+} are stable in the ZnS:Cu:Al phosphor crystal. One might say that the ZnS:Cu:Al green phosphor is the ideal CL phosphor. In practice, however, a-ZnS powders are contaminated with Cl^-, giving rise to the brilliant blue PL and CL. City water usually contains Cl^-, and Cl^- is not completely removed from the water by ion-exchange resins. Because the ZnS:Cu:Al green phosphor is produced from the contaminated a-ZnS powders, it too is contaminated with Cl^- and oxygen, resulting in the variation of the CL color shift and CL intensities. Therefore the use of nonluminescent a-ZnS powder for the production of ZnS phosphors is recommended.

The conclusion is that crystal lattices of phosphor particles are stable under irradiation of energetic particles of electrons and photons (vacuum UV light). This means there is no dislocation of lattice elements by the incident particles. The coloration of phosphor screens in practical CL and PL devices is not caused by the formation of color centers. The coloration is due to adsorption of residual gases in vacuum devices onto the surface of the phosphor particles. In many cases, the coloration (burning) of phosphor screens in vacuum is solved by vacuum technology in PL and CL devices by reducing the residual organic gases. The getters used in the devices do not adsorb the organic gases, and surfaces of phosphor particles adsorb the organic gases. The details are discussed in Chapter 3 and 4.

2.7 Voltage Dependence Curve of CL Intensities

The surface of produced phosphor particles is usually contaminated with something. The contamination generates a nonluminescent layer on the surface of phosphor particles. The measured CL intensities on phosphor screens vary with the contamination level.

It is well-known that the nonluminescent layer (i.e., layer containing surface recombination centers) covers the surface of commercial phosphor particles. To determine the nature of the materials of the nonluminescent layer, the surfaces of phosphor particles can be analyzed by various commercial instruments. For instrumental analysis of the surface, the phosphor powder is placed in a vacuum chamber. The conditions of the chamber differ from the vacuum envelope of CL display devices. Consequently, the measured results [60] do not supply information comparable to the nonluminescent layer of the commercial CL phosphors used.

The nonluminescent layer of commercial CL phosphors is simply detected by observing the PL and determining the voltage dependence curve (VDC) of CL intensities [61]. If ZnS phosphors are produced without flux, and if the produced phosphors in the crucible are immediately placed under high vacuum ($<10^{-6}$ Torr), then the ZnS phosphors emit a brilliant PL under UV irradiation of energy greater than the band gap E_g (3.7 eV = 300 nm). When the vacuum slowly breaks, the PL intensity rapidly decreases to 1/5 of that in vacuum. The PL intensity back again to the original level by a degassing process at 600°C in the pumping vacuum. The PL intensity results indicate that the adsorbed gases from ambient air form a nonluminescent layer on the surface of the ZnS particles produced without flux. The nonluminescent layer from the adsorbed gases can be neglected in practical CL phosphor screens, which are under high vacuum ($<10^{-8}$ Torr) subsequent to the degassing process in CRT production. Therefore, the nonluminescent layer from the adsorbed gases is not of practical concern.

In that the ZnS phosphor is produced without flux, the average particle size is small (<1 μm) for practical use. To obtain a high CL intensity, the commercial ZnS phosphors (4 μm) are usually produced by the addition of Na-halides as the raw materials of the flux [7]. Commercial ZnS phosphors do not emit PL, or emit a faint PL, under UV irradiation shorter than 300 nm, although the phosphors are under high vacuum during the degassing process. The surface of the ZnS phosphor particles is covered with a nonluminescent layer of nonvolatilized material — probably inorganic material. The nonluminescent layer absorbs the incident UV light before reaching the phosphor particles. Further information on the nonluminescent layer cannot be obtained from PL studies. The presence of the nonluminescent inorganic layer on the surface of phosphors can certainly be studied with VDCs.

As electrons enter the phosphor particles, the CL intensities are linear with the energy W of the electrons entering the phosphor particles, and W is expressed by

$$W = k_{10} V_a I_e \tag{2.15}$$

where V_a is the anode voltage, I_e is the electron beam density on the phosphor screen, and k_{10} is a constant. According to Equation (2.15), CL intensities vary with either V_a under a constant I_e or I_e with a constant V_a. Figure 2.18 shows CL intensities with various V_a at constant $I_e = 0.2$ mA (mm spot)$^{-1}$ (Figure 2.18A), and with different I_e at constant $V_a = 20$ kV under a 0.2-μs pulse of 60 Hz (Figure 2.18B). Both curves are obtained using the same energy range

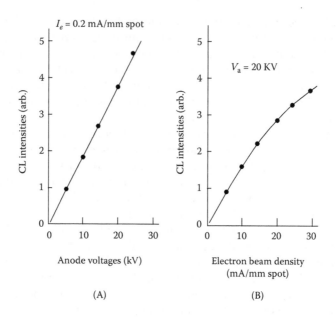

FIGURE 2.18
CL intensities of ZnS:Cu:Al phosphor under electron-beam irradiation as a function of dumped energies on screen: (A) by anode voltages under constant $I_e = 0.2$ mA (mm spot)$^{-1}$, and (B) by electron beam current densities with constant $V_a = 20$ kV.

between 0 to 6 W(mm spot)$^{-1}$ dumped on the phosphor screens. It is obvious from Figure 2.18 that the CL intensities are linear with V_a in the entire energy range, but are sublinear with I_e.

The sublinear dependence of CL intensity with I_e (Figure 2.18B) has been discussed in terms of a nonradiative Auger process at high EH density in phosphor particles [17, 62–64]. The authors derived this conclusion from the I_e-dependence curve; however, they did not measure the VD curve in the same energy density range. Referring to the VD curve in Figure 2.18A, it is argued that the sublinear dependence observed in Figure 2.18B is not related to the nonradiative Auger process. The sublinear dependence with I_e relates to the EH density in the volume, which is determined by V_a; the density of EHs increases with I_e. Some EHs just recently generated might selectively migrate in a path that has been taken by others — that is, correlated migration of EH. This correlated migration prolongs the migration distance of EHs. Because $L_{EH} = \varphi$ for ZnS:Cu:Al phosphors, the prolonged EHs reach the surface of particles, and they nonradiatively recombine at the surface recombination centers. This conclusion is supported by the fact that the deviation from linear is enhanced by small phosphor particles. The results in Figure 2.18 indicate that if one pays attention to a brighter CL phosphor screen, one should take the high V_a rather than the increase in I_e. This is an important conclusion for the development of CL devices; especially a flat CL device such as the FED.

We must know more about the properties of VD curves. The penetration depth of incident electrons into phosphor particles changes with V_a. If the surface of the phosphor particles is covered with a nonluminescent inorganic layer, the incident electrons lose some amount of energy by penetration through the nonluminescent layer. Also, it has been assumed that the threshold voltage V_{th}, at which CL appears, corresponds to the thickness of the nonluminescent layer. The V_{th} of commercial phosphors occurs at around 1000 V, and the CL intensities are linear with V_a between 4 and 10 kV. The linear line intercept of the transverse axis (zero CL intensity) at constant V_a (2 kV) is called the dead voltage V_d. V_{th} differs from V_d. V_d (= 2 kV) is obtained with many phosphor screens, but V_{th} sensitively changes with phosphors and screen structure.

We must determine the V_d of commercial CL phosphors. Fortunately, we can completely clean away the nonluminescent layer on the surface of Y_2O_2S particles (as-produced) by chemical etching in HNO_3 solution (10%) [22]. The etched surface is stabilized by dipping in H_3PO_4 solution (1%). The phosphor screens used for measuring VD curves are made on a conductive substrate by spreading the phosphor particles without overlap (i.e., a single layer with separated particles). An anode potential is applied to the conductive substrate against the cathode. For measuring VD curves of CL, the spot size of the electron beam and the electron beam current (i.e., electron beam density) must remain constant with different anode voltages. Figure 2.19 shows VD curves of the CL from the etched phosphors and as-produced samples at around threshold voltages. The CL of the etched phosphor appears at above 115 V, and CL intensities linearly increase with V_a, according

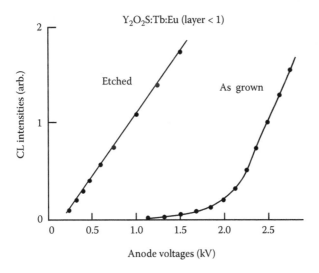

FIGURE 2.19
Voltage dependence curves of the CL intensities of Y_2O_2S:Tb:Eu phosphor particles as produced and etched by HNO_3 solution (10%).

to Equation (2.15). On the other hand, a CL as-produced phosphor has V_{th} = 1 kV and V_d = 2 kV. There is no VD curve between etched and as-produced phosphors (V_a between 115 and 1000 V) under different etching conditions. It might be concluded from the results in Figure 2.19 that V_d is not correlated with the thickness of the nonluminescent inorganic layer.

Nonluminescent materials are electric insulators, which instantly form an electron cloud on the front surface of the insulator when the electron beam enters [21]. The electron cloud shields the insulator, which is interpreted as a negative charge build-up. Incoming electrons must penetrate through the electron cloud to reach phosphor particles, losing energy in the process. To confirm the energy loss resulting from penetration through the electron cloud, SiO_2 microclusters (0.1 wt.% Y_2O_2S) uniformly adhere to the surface of etched Y_2O_2S:Eu phosphor particles. SiO_2 is a typical insulator. The negative field generated by the electron cloud on the SiO_2 microclusters effectively shields the phosphor particles. The resultant VD curve coincides with the VD curve of the as-produced phosphor in Figure 2.19; V_d = 2 kV. Figure 2.20 illustrates a model in which incident electrons penetrate through the electron cloud on a phosphor particle.

Phosphor particles are also insulators, but phosphors are particular insulators that contain recombination centers of EHs. The holes in phosphor particles disappear by recombination of EHs at the luminescence centers. The field generated by the electron cloud on cleaned phosphor particles is much weaker than that on SiO_2. The details of the formation of an electron cloud on phosphor particles are quantitatively described in Section 4.4. When an anode field over a phosphor particle is higher than the negative field of the electron cloud on phosphor particles, the negative field on the phosphor particles is concealed by the anode field. Consequently, the incoming electrons penetrate through the electron cloud without any energy loss, and penetrate into the phosphor particles. This gives V_{th}, which corresponds to the creation energy of EHs in phosphor particles: $3E_g$. This is true for ZnS and ZnO phosphors. The V_{th} of ZnS:Ag:Cl (blue) and ZnS:Cu:Al (green) phosphors is 11 V (= 3 × 3.7 eV), and that for ZnO:Zn (white) is around

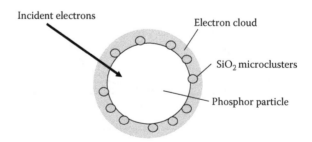

FIGURE 2.20
Schematic illustration that incident electrons penetrate through the electron cloud on SiO_2 microclusters adhered to phosphor particles.

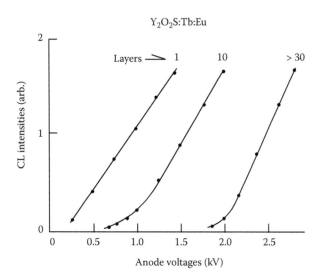

FIGURE 2.21
Voltage dependence curve of etched Y_2O_2S:Tb:Eu phosphor screens in different layers (1, 10, and 30) of phosphor particles (average).

10 V (= 3×3.2 eV). The determined V_{th} values of oxy-compound phosphors, such as Zn_2SiO_4:Mn and Y_2O_2S:Eu (or Tb), are around 110 V, greater than $3E_g$. For example, Zn_2SiO_4 has $E_g = 5.5$ eV [4] ($3E_g = 16.5$ V) and Y_2O_2S has $E_g = 4.7$ eV [5] ($3E_g = 14$ V). The reasons for $V_{th} = 110$ V of some phosphors are not clear. Although V_{th} does not correspond to $3E_g$, the CL of the phosphors linearly increases with V_a above 110 V. The results indicate that $V_d = 2$ kV for commercial CL phosphors relates to electron cloud formation in front of the insulators, and not to the thickness of the nonluminescent layer.

The electron cloud forms on the surface of phosphor particles arranged at the top layer (i.e., the electron gun side) of the screen. The concealment of the electron cloud by an anode field (F_a) is weakened with increasing distance from the anode. When a phosphor screen forms on a conductive substrate that has V_a, F_a is a function of the number of layers (N_L) of phosphor particles on the anode substrate [5], and $F_a = V_a \cdot N_L^{-1}$. Figure 2.21 shows VD curves for etched Y_2O_2S:Tb:Eu phosphor screens with 1, 10, and 30 layers of phosphor particles. The V_{th} of VD curves indeed shifts to higher voltages with increasing thickness of the phosphor screen, and the VD curves of screens greater than 30 layers coincide with the curve of the as-grown phosphor in Figure 2.19, showing that F_a no longer influences the top layer of thick screens.

The difference between the V_{th} and V_d of VD curves has been clarified. Now we discuss the VD curve in practical V_a. Figure 2.22 shows VD curves for the Y_2O_2S:Eu red phosphor up to 26 kV, which is the practical V_a [4]. The CL intensities from etched Y_2O_2S:Eu (clean) phosphor are linear over the entire range of V_a above V_{th} (= 110 V). The surface of commercial Y_2O_2S:Eu

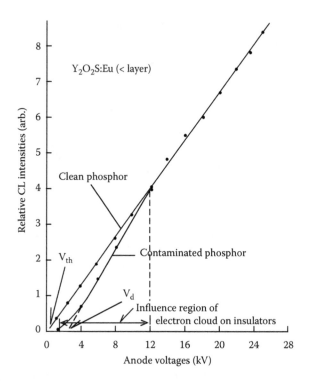

FIGURE 2.22
Voltage dependence curve of Y_2O_2S:Eu phosphor screens of etched phosphor particles by HNO_3 (clean phosphor) and contaminated phosphor particles, explaining the difference between the threshold voltage V_{th} and the dead voltage V_d.

is heavily contaminated with the nonluminescent inorganic materials such as the by-products of the raw materials, pigmentation by Fe_2O_3 and $CaAl_2O_4$, SiO_2 microclusters (surface treatment), and the interface layer. The electron cloud forms on the contaminant materials, giving $V_d . V_a$ between 4 and 12 kV as the range in which incident electrons lose their energy by penetration through the electron cloud. The ratio of the energy loss (E_{ec}) to the energy given to the phosphor (E_p), (E_{ec}/E_p), decreases with increasing anode potential. This gives a large constant k_{10} (steep slope) on the VD curve. Above 12 kV, the (E_{ec}/E_p) is negligibly small, so that the clean and contaminated phosphors have the same VD curve. If the surface of the as-produced phosphor is deliberately coated with a thin film of nonluminescent material, the threshold voltage surely shifts to a high V_a from 2 kV, corresponding to the film thickness of the nonluminescent inorganic material.

As a summary of the study of VD curves, a phosphor screen with less than three layers should be made on the conductive substrate for measuring VD curves. If the CL measurements are made with $V_a < 12$ kV, the CL intensities are markedly influenced by the conditions of surface contamination of CL phosphors. The appropriate V_a (>12 kV) should take into account an evaluation of energy conversion efficiencies of CL phosphors in the development of CL display devices.

2.8   Energy Conversion Efficiencies of CL Phosphors

The determination of the energy conversion efficiencies (η) of CL phosphors requires certain measurement conditions, including (1) particle size (>4 μm) for ZnS phosphors and other phosphors (>1 μm), and (2) anode potential (V_a > 12 kV). Many reported η values fill the conditions for these measurements, and the energy of emitted CL light is determined by thermopiles. Table 2.4 gives the reported η values [4, 31, 32, 65–67].

The η values in Table 2.4 are verified by the number of emitted photons from the CL phosphor screen. Using the ZnS:Cu:Al green phosphor (η = 21%), we can calculate the average number of photons (N_{photon}) from ZnS phosphors per one entering electron. The energy of the light generated in a phosphor particle by a 25-keV electron is $25 \times 10^3 \times 0.21 = 5250$ eV. The average energy of the green band (530-nm peak) is 2.3 eV. Then, N_{photon} is calculated as

$$N_{photon} = 5250 \text{ eV}/2.3 \text{ eV} = 2283 \tag{2.16}$$

CL is a consequence of the radiative recombination of EHs generated in phosphor particles. The number of EHs in the crystal is determined by changes in electric current in single crystals with and without electron irradiation. The creation energy of one EH is $3E_g$. The E_g of cub-ZnS is 3.7 eV [68]. The number of EHs (N_{EH}) generated by 25 keV electrons is

$$N_{EH} = 25 \text{ keV} \cdot (3 \times 3.7 \text{ eV})^{-1} = 2252 \tag{2.17}$$

Assuming that one EH generates one CL photon, the ZnS green phosphor emits 2252 photons per one entered electron. N_{EH} (= 2252) agrees with N_{photon} (2283).

TABLE 2.4

Energy Conversion Efficiencies η of Typical CL Phosphors

Phosphor	η (%)	Ref.
ZnS:Ag:Cl (blue)	23	65
ZnS:Cu:Al (green)	19–21	31, 32
Y_2O_2S:Eu (red)	13	31, 32
Y_2O_3:Eu (red)	8	66
Zn_2SiO_4:Mn	8	67

We confirm the high accuracy of the measurements of η values in Table 2.4 with the calculation of N_{photon}.

The calculations are also applicable to the Y_2O_2S:Eu red phosphor, which has $E_g = 4.7$ eV [5], $\eta = 13\%$, and a photon energy of 2.0 eV (red light at 627 nm). The calculated results are $N_{photon} = 1625$ and $N_{EH} = 1773$, respectively, showing good agreement.

The calculations indicate that the η values of commercial CL phosphors were well-optimized prior to 1975. Since then, a large improvement in the values of commercial blue and green ZnS phosphors and red Y_2O_2S phosphor is not a concern in the study of CL phosphors.

We have calculated N_{photon} from the energy conversion efficiencies (W/W). Recently, many researchers and engineers have used the lumen efficiency (lm/W) as the energy conversion efficiency [69]. The lumen efficiency is uniquely determined from the spectral distribution of the light, and the lumen corresponds to the brightness. However, lumen efficiency (lm/W) is not adequate for a discussion of the energy conversion efficiency of display devices. The brightness of illuminated materials increases with increasing input power with constant energy conversion efficiencies. In practice, the upper limit of input power to the light source is the important concern for the upper brightness (lm). For example, the brightness of tungsten lamps goes up to the upper limit (melting down of tungsten filament), which gives a brightness of 50 to 25,000 lm, depending on the input power with an energy conversion efficiency of 0.8 (W/W). High-brightness light-emitting diodes (HBLEDs) [70] emit photons by the recombination of injected electrons. The efficiency of HBLEDs should be evaluated by quantum efficiency (i.e., number of emitted photons per number of injected electrons). A typical HBLED operates at 60 A cm^{-2} for TV display devices [71]. A current of 1 A contains 1.6×10^{19} electrons, so that 10^{21} electrons are injected into the practical HBLED. ($= 60 \times 1.6 \times 10^{19}$). If one injected electron generates one photon in HBLEDs, then the LED can emit 10^{21} photons cm^{-2}, corresponding to the image luminance of a CRT screen. In reality, the crystal defects in HBLEDs act as nonradiative recombination centers, and about 40% of injected electrons recombine nonradiatively [71]. The nonradiative recombination heats up HBLEDs to around 200°C, which seriously shortens their operational life. The heating temperature determines the upper limitation of luminance in HBLED operation, and it is not the lumen efficiency (lm/W).

Many reports on CL phosphors still focus on an improvement in the energy conversion efficiencies of practical CL phosphors and CL nano-particles [46–50, 72]. For example, a recent article claimed 58% improvement in the CL efficiency of commercial ZnS:Ag:Cl blue phosphors by KOH washing [73]. They measured CL intensities with $V_a < 1000$ V. They actually detected the inorganic residuals on the surface of the ZnS phosphor particles by CL measurements.

Many practical PL screens are excited by the absorption of UV light, corresponding to the charge-transfer band or f → d transitions. The charge-transfer band is formed by electron transfer from the activator ion to surrounding anions, so that the absorption intensities change with activator concentration. The charge-transfer band belongs to the direct excitation of activators, which is less dependent on lattice vibration (heat), so that the PL intensities hold at high-temperature operation. This is an advantage of PL applications such as fluorescent lamps and white-emitting LEDs.

3

Improved Luminance from the Phosphor
Screen in Display Devices

Optimized phosphor powders are screened onto the faceplate to display light images on the screen. Phosphor powder consists of 10^{10} particles per gram, which is comparable to the number of stars in our galaxy. Huge numbers of particles of commercial phosphor powders widely distribute in various sizes and shapes. The CRT (cathode ray tube) industry has used commercial phosphor powders for more than 50 years. This chapter focuses on these phosphor screens in terms of (1) improvements in screen luminance, (2) the lifetime of phosphor screens, and (3) the application of phosphor screens to flat cathodoluminescent (CL) devices.

3.1 Improvements in the Screen Luminance of Color CRTs

The human eye observes light images on phosphor screens. The human eye has adjusted daytime scenes for 7 million years. These scenes are made by sunlight reflected onto various materials. The light consists of photons that have energy; and the number of photons in daytime scenes can be determined from the measurement of that energy. The photon number is $10^{21} \cdot s^{-1}$ cm^{-2} [4]. If the images projected onto the screen in display devices are made by the photon density of $10^{21} \cdot s^{-1}$ cm^{-2}, one can comfortably watch the images on the phosphor screen in illuminated rooms.

In practice, display engineers have not counted the photon numbers from CL phosphor screens, notwithstanding the importance evaluating image luminance on screens in display devices. They measure screen luminance, L_{screen} (cd m^{-2}, in some cases ft-L). The L_{screen} from the phosphor screen in color CRTs has been evaluated with the emitting screen in white (9300°K ± 27 MPCD), that is, white luminance, under the scanning conditions of an electron beam with the NTSC (National Television System Committee). White L_{screen} is determined by the light intensity per unit screen area, and is expressed in units of cd m^{-2}.

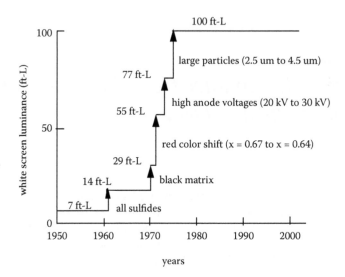

FIGURE 3.1

Stepwise improvement in screen luminance of color CRTs for TV sets, with the reasons. Luminance is given in units of ft-L; 1 ft-L = 330 cd m^{-2}.

Although the phosphor screens in color CRTs was irradiated with the electron beam having 20 keV, L_{screen} in early color CRTs (i.e., prior to 1970) was 25 cd m^{-2}, which was not bright enough to watch images in illuminated living rooms. The TV images were viewed in dimmed rooms, such as movie theaters. The display community looked for more efficient CL phosphors. Contrary to expectations, the L_{screen} has improved to 330 cd·m^{-2} (more than 10 times) by manipulating the CL from phosphor particles and keeping η values [3].

Figure 3.1 shows the stepwise improvement in the L_{screen} of color CRTs, along with the reasons for those improvements. The achieved luminance allows watching TV images in illuminated rooms. The items manipulated included (1) the enlargement of phosphor pixel size by the application of a black matrix (BM), (2) an increase in the lumen-weights in red luminance by shifting the red color to orange (x = 0.64 from x = 0.67), (3) an increase in the energy of electrons irradiated on the phosphor screen by V_a (30 kV from 20 kV), and (4) an increase in ZnS particle size (4 μm from 2 μm).

L_{screen} is measured by the luminance in a unit screen area, and differs from the luminance of the scanning light spot L_{spot} (in mm). For example, L_{screen} = 330 cd m^{-2} screen of phosphor screen is made by the equivalent of L_{spot} = 50,000 cd m^{-2} (peak) [74]. L_{screen} is an important concern, but L_{screen} does not allow a calculation of the photon density emitted from phosphor screens onto which electrons irradiate. The calculation of photon density is a scientifically important concern for the optimization of images on phosphor screens. We can calculate the photon density of the scanning light spot on the phosphor screen in CRTs — that is, the irradiated energy on the phosphor spot (pixel).

A typical color CRT for TV sets operates under the following conditions: $V_a = 30$ kV, $I_e = 1.0$ mA, and spot size = 1.0 mm. The energy of the electron beam $(W = V_a\,I_e)$ is $(30 \times 10^3) \times (1 \times 10^{-3}) = 30$ W per spot. The energy density of the electron beam spot is $30 \times 10^2 = 3$ kW cm^{-2}. With $\eta = 21\%$ for the ZnS:Cu:Al green phosphor (Table 3.3), the energy emitted CL light (W_{CL}) is

$$W_{CL} = 0.21 \times 3 \times 10^3 \text{ W cm}^{-2} = 630 \text{ W cm}^{-2} \tag{3.1}$$

The ZnS:Cu:Al phosphor has a broad CL band in the green wavelength spectral range. When we take the energy of peak wavelength at 530 nm as the average, the energy of green light is 2.3 eV $(= 3.7 \times 10^{-19}$ W). The number of photons N_{spot} emitted from the scanning spot on the phosphor screen is

$$N_{spot} = 630 \times (3.7 \times 10^{-19})^{-1} = 1.8 \times 10^{21} \text{ photons s}^{-1} \text{ cm}^{-2} \tag{3.2}$$

Thus, the peak luminance of 50,000 cd m^{-2} corresponds to 1.8×10^{21} photons s^{-1} cm^{-2}. Images on phosphor screens in current color CRTs for TV display are well optimized for displaying images of daytime scenes on phosphor screens.

3.2 Correlation between Screen Luminance and Spot Luminance

In the above discussion, there are L_{spot} and L_{screen}, which do correlate with each other. Commercial luminous meters measure L_{screen}. These luminous meters detect the time-averaged photons per defined screen area, and then convert it as the time-averaged luminance (cd m^{-2}). Figure 3.2 schematically illustrates L_{spot}, and L_{screen}. Measured area in screen is expressed by green, and red lines represent light intensities emitted from the measured area in screen. According to the scheme in Figure 3.2, we now discuss the measured L_{screen}.

In a CRT, the electron beam scans the entire phosphor screen from left to right and from top to bottom, and each phosphor pixel (corresponding to the beam size) in the screen is sequentially and periodically emitted by the scanning electron beam. L_{screen} is given by the light from a defined screen area (S_{def}) in the screen in unit time (in seconds). Although S_{def} has many emitting lines, the total photon number from S_{def} is proportional to the dumped energy on S_{def}. As the energy of the electron beam, which gives the total screen area (S_T), is W, the dumped energy on S_{def} (scanning lines) is given by W (S_{def}/S_T). A 20-inch screen has $S_T = 1260$ cm^2, $S_{def} = 1$ cm^2, and W = 30 watts. The dumped energy on S_{def} (scanning lines) is 24 mW cm^{-2} (= 30 W/1260 cm^2). If the S_{def} is steadily (dc) irradiated with electrons of energy of 24 mW cm^{-2} $(V_a = 30$ kV, $I_e = 0.8$ μA cm^{-2}), one obtains 330 cd m^{-2}.

L_{spot} is given by a single light intensity emitted from the pixel. The phosphor screen of 1-mm diameter (phosphor pixel) is irradiated with an electron beam

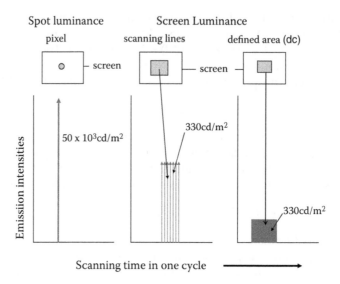

FIGURE 3.2
Schematic explanation of measured spot luminance, and screen luminance by scanning lines and defined area within a refresh cycle. Green areas in the screen (top) represent the measured area, and red lines represent CL emissions measured by the luminous meter.

spot of energy $W = 3 \times 10^3$ watt/spot for 1.7×10^{-7} s. The irradiated energy on the phosphor pixel in unit time (1 s) is $3 \times 10^3 \times 1.7 \times 10^{-7} = 5.7 \times 10^{-4} = 0.57$ mW spot. The area of the spot is calculated as $2\pi r^2 = \pi \varphi^2/2 = 1.6 \times 0.1^2 = 1.6 \times 10^{-2}$ cm². The energy density irradiated of the pixel is $W = 0.57$ mW/$1.6 \times 10^{-2} = 35$ mW cm⁻², which agrees with 24 mW cm⁻² of L_{def} (scan). The calculation shows that if the time-averaged L_{spot} is measured with a phosphor pixel in unit area, the measured L_{spot} will be equal to L_{screen}; that is, L_{spot} L_{screen}. If the measurement of L_{spot} is made in the phosphor pixel, the calculated $L_{spot} = 20 \times 10^3$ cd m⁻² (= 330 cd m⁻² /1.6×10^{-2}). The measured L_{spot} (50,000 cd m⁻²) is determined from the peak intensity of the pulsed emission. If the measurement is made with all emitted photons in the pulse, L_{spot} will decrease to the calculated range. Then, one can say that L_{spot} and L_{screen} are correlated. One can then derive Equation (3.3) from the calculations above:

$$L = \int \int L_0 \, dt \, ds \tag{3.3}$$

where L is the luminance from the phosphor screen, L_0 is the peak luminance, s is the screen area, and t is time. Equation (3.3) can be applied because the luminances are measured by physical means (i.e., photometers).

It is also said that the scientific study of light images on phosphor screens can be performed by determining the number of photons emitted from the screen, L_{spot}. After confirming Equation (3.3), we can use L_{screen} in the evaluation of image luminance of display devices in the quality control of produced CL display devices.

3.3 Image Luminance on Screens Perceived by the Human Eye

Images on phosphor screens in CRTs are generated by rapidly scanning light spots from left to right and top to bottom within refreshing cycles. The human eye, however, perceives planar images on the screen, owing to the effect of after-images on the human eye. Image luminance is given by the definite integral of Equation (3.3) with pixel size(s) and time given by 1/refreshing cycles (t). The minimum refreshing cycle is 30 Hz. Optimized pixel luminance L_{spot} is given by the input power $W = 3 \times 10^3$ watt for 0.17 μs to phosphor pixel, and the L_{spot} is constant for practical display devices. The pixel size on the screen is also constant in the given display device. Variables in Equation (3.3) are irradiation duration t and luminance L_0. We can calculate L_0 for a given L with various t. L_0 corresponds to the input power to the pixel, which markedly changes with pixel size and irradiation time of the electron beam on the pixel, as calculated in Section 3.1.

If the electron beam scans the phosphor screen with noninterlace and with refreshing cycles of 60 Hz, the irradiation time on the phosphor pixel is 0.1 μs. If the horizontal line holds for one scanning line (μs), the holding time is 32 μs, that is, 320 times the t of the point scan. Therefore, the input power on the pixel reduces to 3 kW$\cdot(320)^{-1}$ = 9 W/pixel. For the field scan, the time prolongs to 10 ms with 7 ms returning time, and the input power reduces to 30 mW per pixel (= $3 \times 10^3 \times 10^{-5}$). In reality, the phosphor screen in CRTs cannot scan with line and field scans. Flat panel displays (FPDs), such as FEDs (field emission displays) and VFDs (vacuum fluorescent displays) which have planar electron sources, can operate with line and field scans. In those devices, we can take the refreshing cycles of 30 Hz, and the number of scanning lines reduces to 262 lines from 525 lines. Holding times are 127μs for a horizontal line, and 20 ms for a field scan. The input powers into the pixel are 2.4 W for the line scan and 75 mW for the field scan. Therefore, FEDs and VFDs may have a pixel luminance that is comparable to the image luminance of a CRT screen, with a very low power input on phosphor pixels. Figure 3.3 schematically illustrates the relationship among point, line, and field scans. Table 3.1 gives the calculated results for point, line, and field scans of FPDs. The low input power is a great advantage of FEDs and VFDs.

TABLE 3.1

Difference in Input Power by Image Scanning Mode

Scanning Mode	Input Power (ratio)	Input Power (Watts)
Point scan	1.00	3000
Line scan	$(1270)^{-1}$	2.4
Field scan	$(2 \times 10^4)^{-1}$	1.5×10^{-4}

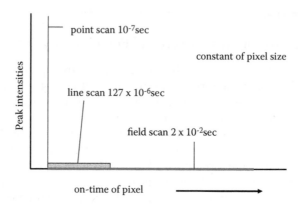

FIGURE 3.3
Schematic illustration of peak luminance *Lp* for point, line, and field scans.

3.4 Lifetime of CL Phosphor Screens in CRTs

Another important concern in practical CRTs is the termination of CL from phosphor screens in CRTs. Termination of CRTs has been attributed to crystal damage to CL phosphor particles under bombardment of electrons and ions. In reality, phosphor screens in CRTs have long lifetimes (greater than 100,000 hours) due to the stability of the crystal lattice and luminescence centers in crystals, as described in Section 2.6. The long lifetimes of phosphor screens is confirmed by recycles of the extinguished CRT with replacement of oxide cathodes and a heater. The phosphor screen that has recycled a few times still emits CL as original luminance. The termination of CRTs is actually determined by the lifetime of the oxide cathodes, including the heater. Although phosphor screens have long lifetimes, there are the disadvantages associated with (1) coloration of the phosphor screen in extinguished CRTs, and (2) a decrease in the CL intensity during operation. We must clarify these problems for production of reliable CRTs.

3.4.1 Coloration of Phosphor Screens by Residual Gases

The phosphor screen in an extinguished CRT has a brown body color. The colored phosphor screen is known as "burning the phosphor screen," and has been attributed to the formation of a color center in the phosphor particles via electron bombardment. According to high-energy science, the energy of an electron beam in a CRT (30 keV) is not large enough to cause displacement of lattice ions of ZnS. The brown body color of phosphor screens in the extinguished CRT must therefore come from something else. Figure 3.4 shows the reflectance spectra of original and burned phosphor screens in a CRT. No absorption band due to the color center is found in the

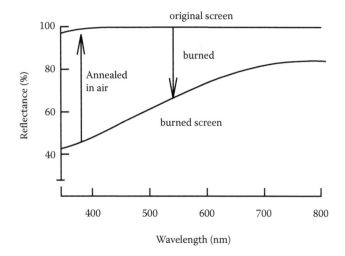

FIGURE 3.4
Reflectance spectra of original, burned, and annealed phosphor screens in a CRT.

spectrum of the burned screen. The burned phosphor screen is bleached by heating at 450°C in air (annealed), but is not bleached in vacuum, thus proving that coloration of the phosphor screen does not result from the color centers. Coloration material should be on the surface of phosphor particles and evaporate to air by burning at 450°C. Such materials are organic compounds.

Solid evidence for coloration by adsorbed organic film on phosphor particles can be obtained from sealed CRTs. If a sealed CRT contains a negligible amount of residual gases ($<10^{-6}$ torr), no coloration of the phosphor screen is observed under electron irradiation, even with 10,000 hours of irradiation. The coloration of phosphor screens only occurs in sealed CRTs, which have residual gases higher than 10^{-4} torr. The residual gases in sealed CRTs can be reduced to an acceptable level by (1) conversion of BaO cathodes and (2) activation of Ba-getters performed in pumping, and as (3) the exhausting glass tube slowly melts down for sealing off the CRT from pumping facilities. The results unmistakably indicate that the coloration of phosphor screens is due to the coloration of the organic materials condensed on the surface of phosphor particles. The origin of these residual gases in sealed CRTs is discussed below.

Phosphor particles can release these organic gases. Host crystals of CL phosphors are compounds in which the cations and anions have large differences in ionic radii; that is, the ionic radii of the anions are much greater than those of the cations. Anions located a few atomic layers from the surface of the compound (crystal boundary) are unstable. These unstable anions are released from the surface volume, leaving anion vacancies. Figure 3.5 illustrates a model in which the surface of a compound crystal releases anions. For example, ZnS forms from Zn^{2+} (0.74 Å) and S^{2-} (1.84 Å). The surface of

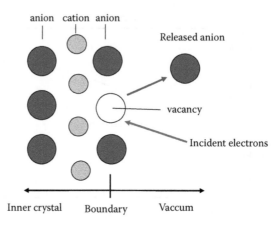

FIGURE 3.5
The surface of compound crystal releases anions by electron irradiation.

the ZnS crystal releases unstable S^{2-}. The vacancies of S^{2-} adsorb the gases from ambient via capillary condensation. In the production process of CRTs, the adsorbed gases are degassed by heating at 450°C under high vacuum ($<10^{-6}$ torr). However, the sealed CRT contains a large amount of these residual gases ($>10^{-3}$ torr). The residual gases do not derive from the phosphor particles.

Modern material science reveals that the materials produced on the Earth always contain these gases, which discharge at meltdown under high vacuum. The materials used in CRTs are not exceptional. The discharged gases from the melted materials can be analyzed by mass spectrometry. The gases are mainly water, methane, carbon oxides, air, etc. In CRT production, we can identify the source of the discharged gases in sealed CRTs.

Fortunately, we can monitor the amount of discharged gases in sealed CRTs by measuring the ion current at grid 1 (G_1) of an electron gun. Figure 3.6 shows the measured results of the sealed CRTs [75]. The CRT is well evacuated to 10^{-7} torr after the degassing process in pumping facilities. When the CRT seals off from the pumping facilities by melting down the exhausting glass tube, the vacuum pressure in the sealed CRT increases to 10^{-5} torr. The increase in pressure results from the trapped gases that are discharged from the melted glass tube, and is independent of the pumping facilities used. By slowly melting the glass tube, some of the discharged gases are removed from the CRT by pumping, and the amount of trapped gases significantly reduces to 10^{-6} torr. If conversion of the Ba-carbonates to BaO cathodes is performed in a sealed CRT (i.e., activation of BaO cathodes), not shown in Figure 3.6, a large amount of carbon oxide gases, as well as other gases, release to the sealed CRT, and the pressure increases to 10^{-2} torr. If conversion of the BaO cathodes occurs by pumping, then the generated gases are immediately sucked out of the CRT. Then, one can relax the residual gases of the conversion of BaO cathodes. The gases are also generated as the raw materials

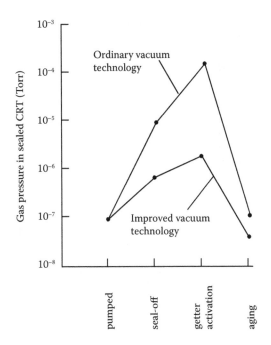

FIGURE 3.6
Changes in vacuum pressure of sealed 0.5-inch CRT by ordinary CRT production process and by improved vacuum technology.

of Ba-getter meltdown prior to evaporation of Ba metal on the glass wall (activation of Ba-getters). The pressure increases to 10^{-4} torr. If the raw materials of Ba-getters are partially activated (~40%) by preheating during pumping, the generated gases are pumped out of the CRT. And then the Ba-getters are activated in the sealed CRT, and the residual gases markedly reduce to an acceptable level. Evaporated Ba-metal films are chemically reactive materials with oxygen, water, and hydrogen gases, and form BaO, $Ba(OH)_2$, and BaH_2 (via chemisorption). The formed Ba compounds are solids with a white body color. The residual inorganic gases in sealed CRTs condense in the solids, reducing the gas pressure of the inorganic gases in the sealed CRT. The Ba-getters do not absorb the self-generated gases in the CRT, although the sealed CRT holds for a week at room temperature, indicating that the self-generated gases are not inorganic gases — they are organic gases.

Phosphor screens in sealed CRTs have a pumping action to the organic gases, under irradiation of a scanning electron beam. Figure 3.7 shows the pumping action of methane (CH_4) gas by a phosphor screen. Methane gas maintains a constant pressure with the phosphor screen alone. For the adsorption of methane gas, the phosphor screen must periodically undergo irradiation by electrons. Adsorbed organic materials partially decompose under bombardment of negatively charged ions, thus giving rise to a light

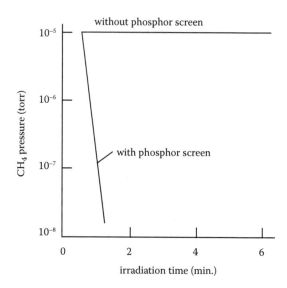

FIGURE 3.7
Decrease in partial pressure of methane (CH_4) gas in sealed CRT under electron-beam irradiation on phosphor screen and without phosphor screen.

brown coloration. Consequently, the phosphor screen is colored by a partially decomposed organic film on the phosphor particles. The coloration of phosphor screen in CRTs is markedly reduced by the application of an appropriate vacuum technology to CRT production.

3.4.2 Decrease in CL Intensity during Operation

In some situations, the CL intensities of the phosphor screen in CRTs decrease with increasing operation time. As described above, the surface of the phosphor particles in the screen adsorbs organic residual gases, thereby reducing CL intensities with operation. The problem can be solved by applying high vacuum knowledge to the CRT production process. In other cases, the phosphor crystals release anions from the surface volume, as illustrated in Figure 3.5. The release of anions involves a shallow volume — a few atomic layers. Because a CRT operates at anode voltages greater than 15 kV, the problem will be concealed in measurement error of CL intensities (see Figure 2.22).

Some CL engineers attribute the decrease in CL to the accumulation of charge of irradiated electrons (Coulomb effect) [16, 17, 76, 77]. The mechanism of the Coulomb effect is obscure. The problems relate to vacuum technology and the application of improper anode voltages. The application of high vacuum technology and V_a greater than 15 kV solves the problem.

There are reports that nonluminescent materials (e.g., ZnO and $ZnSO_4$) are formed on the surface of commercial ZnS phosphors for VFDs and FEDs [72, 78–82]. These experiments were performed in a vacuum chamber in which the oxygen gas ($>10^{-5}$ torr) is deliberately introduced. In CRTs, the getters effectively and selectively absorb oxygen gas to pressure levels less than 10^{-8} torr; that is, much less than 10^{-5} torr. These results are inapplicable to practical CRTs.

3.4.3 The Lifetime of Oxide Cathodes

As described, phosphor screens in CRTs are not damaged by electron-beam irradiation. The termination of CRTs is determined by the lifetime of the BaO cathodes, which emit thermoelectrons. BaO cathodes are very tough materials and emit thermoelectrons at a vacuum pressure of approximately 10^{-2} torr for awhile. If the phosphor screen quickly pumps out the residual gases in the sealed CRT to a level less than 10^{-8} torr, the BaO –cathode serves as the electron source in CRTs. This is one reason why manufactured CRTs require aging (i.e., operation of CRTs) prior to shipping. During the aging process, the phosphor screen adsorbs the organic residual gases. The oxide cathodes are also damaged by these residual gases.

According to probability theory, the lifetime can be expressed by the holding time (life). Survivors of the lifetime (SR) can be expressed by

$$SR = A\,e^{-\lambda t} \tag{3.4}$$

where λ is the probability constant, t is time, and A is a constant. The lifetime curve should be plotted on semi-log graph paper as log (SR) versus t; log (SR) = A − λt. If n-controlling factors ($\lambda_1, \lambda_2, \dots, \lambda_n$) are involved in the lifetime, the lifetime curve consists of n straight lines in the different operating times. We can identify each controlling factor of the lifetime from the curve.

As shown in Figure 3.8, two factors (absorption of residual gas and ion bombardment) are involved in the lifetime of BaO cathodes. The absorption

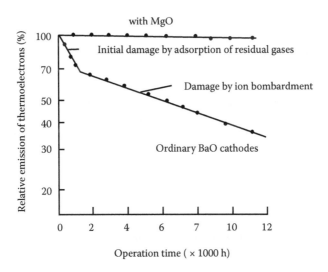

FIGURE 3.8
Lifetime curves of electron emission from oxide cathode in the CRT.

of the residual gases dominates in initial damage up to 1000 hours, and damage by ion bombardment dominates during subsequent operating time. At 1000 hours of operation, the vacuum pressure in CRTs lowers to a level below 10^{-8} torr as a result of the pumping action of the phosphor screen. The damage to BaO cathodes is negligible in a vacuum of 10^{-8} torr. If improved vacuum technology applies to CRT production, then the time for initial damage significantly shortens the lifetime curve. We cannot completely eliminate residual gases in a sealed CRT. Thus, protection against damage by the bombardment of ionized residual gases must be considered.

This can be achieved by the addition of a small amount of MgO (or RE oxides) to BaO cathodes (see curve with MgO in Figure 3.8). MgO, which has a high melting temperature, can protect BaO cathodes from direct ion bombardment. Consequently, the lifetime curves of CRTs may hold the initial luminance for 12,000 hours in Figure 3.8, and will extend further in time, at which the heating of BaO cathodes determines the lifetime of thermoelectron emission (>30,000 hours).

The results in Figure 3.8 provide important information for the evaluation of the lifetimes of display devices. If there is no factor that determines the initial damage, the lifetime curve maintains the initial luminance up to the time that some lifetime-influencing factor becomes effective during operation. The screen luminance of commercial LCDs and PDPs decreases with operational hours right from the start of operation, showing that these display devices possess initial damage factors. The nominal lifetimes of LCDs and PDPs are estimated from the time that gives half the luminance of the initial values. The actual lifetime will be shorter than the estimated lifetime. By plotting the lifetime data according to Equation (3.4), one can determine the initial damage factor.

3.4.4   The Lifetime of PL Devices

Phosphor screens are used in PL devices, such as fluorescent lamps and PDPs, in which phosphor screens are placed in vacuum. One can obtain the lifetime curve of PL intensities as shown in Figure 3.8. The electrons in CL devices have much higher energy than PL photons. One can therefore say that the host crystals and activators are stable in PL operation in a vacuum. The decrease in PL intensities of PL devices is attributable to the absorption of residual gases in those devices.

3.5   Flat CL Displays

Phosphor screens in CRTs have the distinct advantage of displaying images equivalent to daytime scenes. Disadvantages of CRTs include the bulky and heavy glass envelope of the vacuum space. The electron beam is deflected by a magnetic coil for scanning the electron beam on phosphor screens. Table 3.2 gives the power consumption of a commercial 19-inch TV set

TABLE 3.2

Power Consumption of 19-inch CRT (100 Watt) as Example

Part	Consumed Power (W)	Consumed Power (%)
Electron beam ($V_a \cdot I_e$)	$27 \times 10^3 \times 1 \times 10^3 = 27$	27
Cathode heater	$6.3 \times 600 = 4$	4
Deflection coil	39	39
Electronic circuits	30	30
Total	100	100

(100 W) and parts: electron beam, cathode heater, deflection coil, and electronic circuits for video and audio. Recent CRTs exhibit a reduction in the diameter of the neck tube, which sheathes the electron gun, resulting in a reduction in power consumption [83] from 65 to 39%. The deflection coil still consumes the most power (39%) in a CRT device. If the deflection space is removed from the CRT envelope, the deflection coil and the shadow mask become unnecessary. Then, the power consumption significantly decreases to less than 50% (half) (e.g., less than 50 W). If a 20-inch LCD panel operates with field scan, the required input power of the backlight is calculated as 360 W. Flat CL devices have a great advantage over LCDs and PDPs in terms of power consumption (one seventh that of LCDs). To remove the deflection coil and shadow mask from CRTs, a two-dimensional planar electron source — not a single point source — is an absolute necessity for flat CL display devices.

3.5.1 Vacuum Fluorescent Display (VFD)

One of the first commercial flat CL display devices was the monochrome VFD [84]; it has multiple cathode filaments parallel to each other, and electrons from cathode filaments uniformly shower (30 m $\cdot$ cm^{-2}) the entire display area, without focusing. No space for electron deflection is necessary in VFDs, thus rendering a flat and thin CL display device. Figure 3.9 shows a commercial VFD.

A VFD displays letters (and numerics), each letter being consisting of several segments (pixels) of phosphor screen on anode electrodes. Electron irradiation of each phosphor pixel is controlled by the potential of the x-y matrix anode electrodes, which are addressed by an electronic chip (on-off). When the anode has a positive potential versus the cathode filaments, electrons in the shower reach the phosphor pixels, and the phosphor pixel emits CL (on). When the anode pixel has a negative potential versus the cathode, the phosphor pixel is shielded by the negative field of the anode. Incoming electrons repulse the negative field and do not reach the phosphor pixel (off). The electrons repulsed by a negative field (stray electrons) reach the positive anodes at the nearest neighbor pixels, giving rise to a brighter edge CL on the emitting phosphor pixel. If the size of the phosphor screen is smaller than the area of the anode electrode, the naked area of the positive anode collects the stray electrons, thus eliminating the edge CL.

FIGURE 3.9
Typical VFD. (*Source:* Reproduced from Internet site of Futaba Electric Co.)

Images on the screen are observed in reflection mode (irradiated side). Referring to Equation (2.2) and Figure 2.3 in Section 2.2, seven layers provide the optimum thickness. For effective control of the incoming electron shower, the phosphor screens must have an appreciable number of holes. The screens of VFDs are produced in two ways: (1) spraying a phosphor slurry, the screen being composed of clumped particles; and (2) a print technique involving an appropriate amount of solvent that has a low evaporation temperature (the solvent makes small through-holes in the screen via a quick-dry technique).

Operation of the chip limits the anode potential to lower than ± 50 V. ZnO CL phosphors ($\eta = 8\%$) emit bluish-white CL under electron irradiation of energy above 10 V ($3E_g = 3 \times 3.2$ eV). Typical operating conditions for a practical VFD are $V_a = 26$ V and $I_e = 0.8$ mA cm^{-2}, giving an electron shower of 30 mW cm^{-2} [85, 86]. The electron shower generates a luminance of 330 cd·m^{-2} per phosphor pixel, which is bright enough for observing the CL letters in illuminated rooms. Irradiation of an electron shower on phosphor pixels is periodically refreshed for the display of different letters without any perception of flicker. The refreshing cycle for displaying letters is empirically determined as 50 Hz, 20 ms for one cycle. Then the VFD engineers define a duty factor (t_{off}/t_{on}). Under a given electron shower, the pixel luminance varies linearly with the duty factor. A high luminance is obtained with a low duty factor. A typical duty factor for a practical VFD is 1/16 — that is, $t_{on} = 18.8$ ms and $t_{off} = 1.2$ ms.

We must understand the high luminance (330 cd·m^{-2}) and low input power for still images in the development of flat CL display devices. As described in Section 3.3, the perceived luminance by the human eye is given by the definite integral of Equation (3.3) ($\iint B_p \, dt \, ds$) with time (t) and emitting area (s).

In a given display, the pixel area (s) is constant and time is limited by the effect of the after-images of the human eye (30 Hz).

Using Equation (3.3), one can calculate a suitable irradiation density for the electron shower on phosphor screens in VFDs from the irradiation conditions of the electron beam of the CRT. In CRTs, electron beams having an energy of 3×10^3 W cm^{-2} scan the entire phosphor screen with NTSC conditions (point scan). The duration of electron beam irradiation on a phosphor pixel is 1×10^{-7} s. When the pixel is irradiated for one frame period (field scan), the irradiation time is 20 ms (2×10^5 times that of the point scan). The power irradiated on the phosphor pixel is 15 mW cm^2 {= $3 \times 10^3 \times$ ($2 \times 10^5)^{-1}$}. The η value of ZnO is 0.08; that is, 1/2.5 that of the ZnS phosphor ($\eta = 0.2$). The shower density of electrons, calculated from CRT operation, is 37.5 mW cm^{-2} (= $15 \times 10^{-3} \times 2.5$), which agrees with 30 mW cm^{-2} for practical VFDs.

For displaying video images on a VFD screen, the size of the dotted phosphor pixel should be less than 0.01 cm^2 (0.1×0.1 cm^2) (one tenth of letter pixel in 0.1 cm^2). The power of dotted phosphor screens in VFDs should be greater than 300 mW cm^{-2}. This is a reason that VFDs cannot display video images on phosphor screens.

Operation of VFDs is the field scan using a short-decay ZnO phosphor. Calculations using Equation (3.3) show an important concern in the development of flat CL display devices. A practical CL flat device will be made by the field scan, with the CL phosphor having a short decay time.

Monochrome VFDs with low V_a (<50 V) have a great advantage in terms of lifetime insofar as the phosphor screen is made from the ZnO CL phosphor. The ZnO CL phosphor is produced by heating ZnO powder in reducing atmosphere (N$_2$ with 5% H$_2$). The luminescence center is assigned an oxygen vacancy, despite the Zn vacancy [7]. In phosphor production, sufficient amounts of oxygen have been removed from the surface of the ZnO CL phosphor particles. Consequently, the ZnO CL phosphor does not release O^{2-} into vacuum via the irradiation of electrons. If a produced VFD exhibits a short lifetime of CL intensity, the short lifetime is not attributed to the ZnO CL phosphor. The short lifetime of the VFD is caused by poor vacuum technology [87]. The problem with the ZnO phosphor is the small size of the particles. As described in Section 2.5, the large ZnO phosphor particles give brighter CL. The production of large ZnO particles remains a future study area for brighter VFDs.

Because color VFDs operate at low V_a (<150 V) with the limitation of the driving chip, commercial color CL phosphors for CRTs cannot be applicable to VFDs. Color VFDs use ZnS phosphors, the production of which differs from that of commercial CL phosphors [88]. If ZnS phosphors are produced without the flux, the phosphor emits CL with V_a greater than 10 V. However, the particles size is small (1 μm), and CL intensity is too low for practical use. A brighter ZnS phosphor for VFD is produced as follows. The ZnS phosphor is first produced with the flux to the proper size (4 μm). After washing and etching the ZnS phosphor with a dilute acid solution, the phosphor, without any addition, is heated again at 950°C under H_2S gas flow for a long time (>5 hours) in a shallow crucible to modify the surface conditions. The resultant ZnS color phosphors emit CL at V_a less than 100 V.

The adhesion of In_2O_3 powder onto the surface of ZnS phosphor particles helps the low-voltage ZnS phosphor for VFDs [89], presumably through surface conduction of the phosphor particles, but this remains unsubstantiated. According to the assumption, the fine metal powder, such as Ag, Cu, Ni, and Co, adheres to the ZnS phosphor particles. Negative results were obtained with the metal powders. The study of the adhesion of In_2O_3 powder is not due to the surface conductance, and must be due to something else.

The ZnS phosphor for VFDs inevitably releases anions to vacuum under electron irradiation (as described in Section 3.4.2), and the residual gases (especially water) are selectively condensed in the holes of anion vacancies via capillary condensation. The ZnS color phosphors for VFDs have the disadvantage of short lifetimes at V_a less than 12 kV. For a long CL intensity lifetime, the ZnS color phosphor should operate at V_a greater than 12 kV.

An advantage of monochrome VFDs is that heating the phosphor screen, by the low-energy density of the electrons dumped on the phosphor screen, occurs only at a negligible level (30 mW cm^{-2}). The screen luminance increases with the energy density of the electrons. If the desired screen luminance is 33,000 cd m^{-2}, the phosphor screen should be irradiated by electrons of energy density 3 W cm^{-2}, (e.g., 60 W for 20-cm^2 VFDs). Then, the thermal quenching of the ZnO phosphor, as well as heating the VFD, becomes a serious problem during practical use.

3.5.2   Horizontal Address Vertical Deflection (HAVD)

Miyazaki and Sakamoto [90] developed the next flat CL display, made by vertical deflection of line electrons from a single filament on the phosphor screen. The device is called horizontal address vertical deflection (HAVD). The thermoelectrons are accelerated by high V_a (>5 kV), and the accelerated electrons in the line are deflected by a pair of electrostatic electrodes (thin plate). The pixels of the phosphor screen are addressed by a vertically deflected electron beam, which is subsequently controlled by the array of grid electrodes in front of the filament. The modulation of electrons by the grid array electrodes allows the high V_a (>10 kV).

Figure 3.10 shows a photograph of the video image on a ZnO phosphor screen (3-inch diagonal) in experimental HAVD. Deflection of the electron beam is made by a ±350-V saw wave to the deflection electrodes. Video signals are supplied to the grid electrode arrayed in line (5 V maximum) above the cathode. The luminance of 230 cd·m^{-2} was obtained with V_a = 5000 V and I_e = 70 μA. However, the electrons from the single cathode filament are insufficient to display images on phosphor screens in HAVD. HAVD never became commercially viable.

FIGURE 3.10
Photopicture of image on the screen of a CL device that is horizontal address vertical deflection. (*Source:* Courtesy of Professor Y. Sakamoto.)

3.5.3 Modulation Deflection Screen (MDS)

Watanabe and Nonomiya [91, 92] invented a 10-inch, color, flat CL display with many cathode filaments. The display is known as the modulation deflection screen (MDS). The fundamental structure of the MDS is very complicated and requires high assembly precision for production. The number of image pixels developed is 228 × 448 triplets. A triplet is a combination of blue, green, and red pixels. The white screen luminance of 680 cd m^{-2} is obtained at V_a = 12 kV. Figure 3.11 shows a photograph of the color image on the phosphor screen. The image quality on the 10-inch color MDS was comparable to that of a 10-inch color CRT screen. The disadvantage of the MDS involves the densely arranged cathode filaments that heat up the MDS, and thus the MDS soon disappeared from the marketplace.

The development of HAVD and the MDS proved that it was possible to make a practical color CL flat display, operating at V_a greater than 15 kV and I_e greater than 10 μA cm^{-2}, and also proved the application of commercial CL phosphors of size 4 μm. Electron showers from cathode filaments are unsuitable for flat CL displays because of the heat of the device. The development of an array of an electron source as flat planar, not electron shower, awaits further study in the area of flat color CL display devices.

FIGURE 3.11
Video image on matrix drive and deflection screen. (*Source:* Courtesy of Drs. M. Watanabe and K. Nonomura.)

3.5.4 Field Emission Display (FED)

Modern microfabrication technology as developed by the solid-state device industry allows for the flat planar array of the electron source (field emitter array, FEA) [93]. Normally, electrons cannot escape from a metal plate unless the metal is heated. It has been known since 1928 that when a high electric field is applied to a metal, the energy barrier between the metal and the vacuum decreases, and the metal emits electrons to vacuum (Flower-Nordheim tunneling) [94]. High electric fields can be applied at distinct points of materials. If an electric field of 10^7 V cm^{-1} is applied to the tip of W, Mo, or Si, which have a work function of about 4.5 eV, the tip emits electrons to vacuum (Spindt-type field emission, FE) [95].

A small FE (μm size) is constructed with an emitter and a gate (grid) that controls electron emission from the FE. A small number of electrons (10^{-9} A) is stably extracted from the FE. An appropriate number of electrons is obtained from an FE pixel that is composed of multiple FEs, and electrons in each FE pixel are focused by the electrode [96, 97]. If the size of the FE pixel corresponds to the size of the phosphor pixel in the screen, the phosphor pixels are addressed by controlling the electrons from the corresponding FE pixels. There is no deflection electrode for addressing the phosphor pixels. The FE pixels and focus electrodes are thin films (in the micrometer range), and they are arranged on a planar substrate known as a field emitter array (FEA). Then, it is possible to produce a flat CL display device (FED) with a combination FEA and CL phosphor screen.

In 1994, Pixel International formed the FED Alliance with Texas Instruments, Raytheon, Motorola, and Fataba. They produced an evaluation sample of a color FED. They produced a narrow gap between the FEA and the phosphor screen (<1 mm) using color VFD phosphors (V_a < 150 V). Their work misled others who were developing FEDs.

With the narrow gap between the phosphor screen (anode) and the FEA, they selected a low V_a to avoid the influence of the anode field on the gate electrode. Their FEDs displayed color still images, but the video images were dimmed with an insufficient input energy [98, 99], as found for color VFDs. The dark images of FEDs were attributed to the low η value of the phosphors. As described in Sections 2.7 and 2.8, the η values of CL phosphors were optimized in the 1970s, and the phosphor screens in FEDs must be irradiated by electrons having energy greater than 12 keV, rather than increasing the number of electrons. The FED Alliance disappeared in the late 1990s. Accordingly, the activity of FEDs ceased in the United States and in Europe. FED activity, however, did move to Asian countries.

The narrow gap of FEDs was a design mistake. The concept of the narrow gap may have come from the structure of the LCD. The LCD is an optical filter and requires a brighter backlight in order to display an image luminance of 10^{21} photons·s^{-1} cm^{-2}. LCDs are power-hungry display devices. The total energy conversion efficiency is less than 1%, and 99% of the input power converts to heat. The brighter LCD devices at present have thicknesses greater than 5 cm for cooling the generated heat from backlight and the device. This means that a narrow gap in the FED is not an essential requirement for the development of FEDs. Another concern is a misunderstanding about CL. CL refers to photons that have energy. For obtaining brighter images such as in a CRT, the energy on the phosphor screen in an FED must be the same energy as for the CRT screen to give a photon number of 1×10^{21}·s^{-1} cm^{-2} in the period of one frame. This concept was neglected by the FED Alliance. Recently, Canon and Toshiba announced their developed FED with V_a = 12 kV and a low electron density [100]. Similar images will be obtained with other FEDs with V_a > 12 kV and a low I_e employing the field or line scan.

Because an FE can be made by either point-to-plane or plane-on-plane geometries [93], different types of FEAs have been proposed. They include the (1) Spindt-type chip [95], (2) metal-insulator-metal (MIM) [101], (3) metal-insulator-semiconductor (MIS) [102], (4) surface conduction emitter (SCE) [100], (5) carbon nanotube (CNT) [103], and (6) ballistic silicon diode (BSD) [104]. From an operational viewpoint of FEDs, the FEA is an electron source for the FED, but a detailed discussion of FEAs is beyond the scope of this book. The proposed FEAs will be evaluated practically in terms of (1) a suitable amount of electron emission for the phosphor pixel, (2) the stability of electron emission for long operation, (3) the size of the planar FEA, (4) the ease of production, and (5) the production cost. Using any FEA, the phosphor pixels should be irradiated with electrons that have sufficient energy to generate the photon number in the period of a single frame scan.

TABLE 3.3

Calculated Operation Conditions of Phosphor Pixel in a Practical FED

Scanning Mode	Irradiated Energy Density (W cm^{-2})	Minimum Anode Potential (kV)	Minimum Electrons per Pixel (1 mm^2)
Field	0.008	>12	>7 nA
Line	1.2	>12	>1 μA
Point	1500	>12	>1 mA

In the development of a reliable FED, one must be sure that the phosphor screens are made by CL phosphors (4 μm as average) for color CRTs. Nevertheless, the nanoparticles emit CL [46–51]; however, the CL intensity from nanoparticles is far from being accepted for practical use (as described in Section 2.5). Thus, we can set the irradiation conditions of electrons on phosphor pixels in FEDs. They are that, at a minimum, V_a > 12 kV, and irradiated energy densities per phosphor pixel are greater than 8 mW cm^{-2} by field scan, 1.2 W cm^{-2} by line scan, and 1500 W cm^{-2} by point scan. If V_a = 12 kV, the number of electrons from an FE pixel (I_e) is calculated as I_e > 0.7 μA cm^{-2} (= 7 μA mm^{-2}) by field scan, I_e > 100 μA cm^{-2} (= 1 μA mm^{-2}) by line scan, and 1 A cm^{-2} by point scan. Table 3.3 gives the calculated results of FEDs, which may display video images comparable to that of CRTs.

CRTs only operate by point scan and avoid the heat-up of phosphor pixels under electron bombardment of high energy density (1.5 × 10^3 W cm^{-2}). Flat CL displays can operate viaboth line and field scan. The field scan of flat displays is only allowed using short decay phosphors (<500 μs). All commercial CL phosphors for CRTs have a short decay (100 μs for blue and green phosphors, and 500 μs for red phosphors). An FED operated by field scan has the distinct advantage of extremely low power consumption — as compared with LCDs and PDPs.

In evaluating the advantages of FEDs over LCDs and PDPs, one must consider LCD operation from the viewpoint of energy consumption. An LCD is an optical filter with 8% transmission (including color filters). In addition, an LCD requires a backlight. The energy conversion efficiency of the backlight, from the input power to the light, is 20% maximum. The total energy conversion efficiency of an LCD is 1.6% maximum — that is, one-tenth that of an FED. By a simple calculation, the input power to the backlight for a 20-inch LCD panel is 3.5 × 10^3 watts to display equivalent images with a CRT. As compared with the input power of a 20-inch CRT screen (100 watt), the LCD consumes 35 times the power of a CRT. Figure 3.12 shows the matrix connections of the FED and LCD. The LCD needs TFT for operation, while the FED does not have it. The FED will save power as well as the cost of TFT. The above calculation shows that the LCDs cannot compete with FEDs with regard to image quality, power consumption, and production cost. Unfortunately, the FEDs developed to date were designed using the wrong conditions.

Great difficulty exists in developing a practical FED — that is, the vacuum technology, rather than the phosphor. CRTs and VFDs use Ba-oxide cathodes,

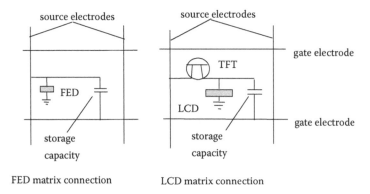

FIGURE 3.12
Matrix connections for FED and LCD pixels.

which function at vacuum 10^{-2} torr for the emission of thermoelectrons. By the pumping action of the phosphor screen, residual gases quickly reduce to a level below 10^{-8} torr, at which oxide cathodes can safely operate. The FEs in FEAs cannot operate at poor vacuum pressures above 10^{-6} torr, as arc discharge occurs between the emitter and gate electrodes. Arc discharge damages FE emitters. The emitters in the FE exist in tiny holes, small cracks, or narrow gaps, in which ambient gases are condensed by the capillary condensation. The condensed gases are not smoothly degassed from the holes and gaps under ordinary degassing process by heat (at 450°C) that is applied to CRT and VFD production. The degassing of the condensed gases in the tiny gaps, especially carbon nanotubes, in FEDs will be a clue for the future development of reliable FEDs.

4

Improving the Image Quality of Phosphor Screens in CL Displays

In cathodoluminescent (CL) display devices, light images are produced by tiny emitted elements (pixels), and image quality is directly related to CL generation. Chapter 3 described various advances in phosphor screens formed from commercial phosphor powders. CL screens inevitably exhibit flicker of images in small sizes that is not perceived by image viewers who watch images at a distance of five times the screen size; (e.g., 3 m). Furthermore, CL screens sometimes show the moiré effect. As a consequence of the brighter images on a phosphor screen, which are equivalent to daytime scenes, TV displays penetrate the illuminated living rooms of families. After 1980, personal computers (PCs) became prevalent on individual desktops, and PC images on phosphor screens are watched at the distance of distinct vision of the human eye (e.g., 30 cm away from the screen). By observing images at a short distance from the screen, the screen luminance of the PC decreases to the level of 150 cd m^{-2}, that is, half that of TV screens, resulting in that the moiré disappears from the CL phosphor screen. Then, image quality on CL phosphor screens, including flicker, smear, contrast, and color fidelity, is a serious problem for protecting the human eye from permanent damage. The image quality of CL phosphor screens relates to (1) the structure of the phosphor screen, and (2) the electrical and (3) physical properties of bulk of phosphor particles in the screens. Chapter 4 discusses these problems, which are definitely present in CL devices, for further improvement of the image quality of CL phosphor screens.

4.1   Threshold Resolution Power of the Human Eye

The human eye sensitively detects and identifies flicker and smeared images even if the images are in a corner of the visual field; this is the consequence of adjustment to the wild life for 700 million years. The perception of flicker in images on phosphor screens will be discussed here.

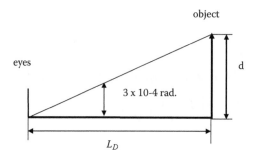

FIGURE 4.1
Schematic illustration of the threshold resolution power of the human eye.

The human eye sensitively detects fluctuating objects in the visual field. Periodically fluctuating objects and lights are perceived as flickers, which irritate the eyes, although the fluctuation is small and is but a corner of the visual field. If the fluctuating distance d (object size) is below the resolution power of the human eye, one might not perceive the flicker. The resolution power is defined by the vision angle, which is 3×10^{-4} radian for the average person, and d changes with distance from the object. Figure 4.1 illustrates the resolution power and d as a function of the distance (L_D) from the screen.

If a viewer watches images on a CL screen at L_D, the minimum resolution (d) on the screen is expressed as

$$d = 3 \times 10^{-4} L_D \qquad (4.1)$$

If $L_D = 3$ m, which is the distance away from the screen of ordinary TV viewers, then $d = 0.9$ mm. If $L_D = 0.3$ m, which is distance of distinct vision, then $d = 0.09$ mm. The CL phosphor screens, which are prepared from commercial phosphor powders, exhibit fluctuations greater than 0.5 mm. Image viewers who are 3 m away from the screen do not perceive the flicker of images, but surely do perceive the fluctuation of images at the distance of distinct vision. We therefore must determine the origin of the fluctuation of the images on phosphor screens.

4.2 Survey of the Origin of Flickers

The human eye perceives two kinds of flickers: (1) one fluctuates with light intensity over time at the same place, $f(t)$; and (2) the other is a fluctuation of light position (d) over time, $f(d, t)$ [105]. Thus, $f(t)$ relates to refreshing cycles of the image frame (i.e., the vertical frequency of the frame), and the study of $f(t)$ is well established by the movie and TV broadcasting industry

(>30 Hz). The empirical findings of the $f(d, t)$ of images on CL phosphor screens at the distance of distinct vision date back to the early 1980s. The perception of image flicker on the CL screens of PCs (i.e., video display terminals, VDTs) changes with (1) phosphors having different decay times of CL under the constant vertical frequency and (2) vertical frequencies of the frame with the same phosphor screen [106, 107]. There thus arises confusion and inconsistency with regard to ergonomics and incorrect assignment of phosphors. Such changes derive from the physical properties of bulk CL phosphor particles under electron-beam irradiation [23]. The flicker of images on CL phosphor screens generates from the perturbation of electron trajectory with the negative charges on phosphor particles. There was, however, no report on the subject prior to 1999. It was thought that the flicker of images on CL phosphor screen derived from $f(t)$. As a consequence, the flicker of images on CL phosphor screens have been analyzed by the modulation transfer function (MTF) (or optical transfer function (OTF)) [108].

The MTF was developed as a new tool for analyzing images from medical diagnosis techniques, photography, printing, and optical lenses, using a Fourier analysis that is a convolution of sine and cosine functions. The MTF is a powerful tool for the analysis of smeared images. The commercial MTF device can detect the $f(d, t)$ of images on CL phosphor screens. The detected fluctuation of $f(d, t)$ coincided with the frequencies of the refresh frequencies. Then one must consider that the flicker by $f(d, t)$ is caused by "mis-landing" of the electron beam on the phosphor pixel (target accuracy).

Target accuracy is a wrong assignment. Using a given CRT, the flicker of images on the phosphor screen is significantly reduced by (1) an increase in the vertical frame frequencies, and (2) a decrease in the electron beam current. If the fluctuation is from target accuracy by deflection yoke (DY) coil or fluctuation of the terrestrial magnetism, the target accuracy of the electron beam does not change with the frame frequency and the amount of the electron beam.

In reality, one does not perceive flicker images with a scanning electron beam of 100 µA, but one surely does perceive flicker images with a 500-µA electron beam. This is one reason that the moiré on TV screens disappears from the CL screen of VTDs, for which the screen luminance (150 cd m²) is half that of TV screens (330 cd m²). The moiré is a large flicker of images that appears on the entire phosphor screen. The disappearance of moiré from phosphor screens in VTDs indicates that moiré is not caused by interference between the aperture size of the shadow mask and the electron beam. A further confusion derives from the irradiation duration of the electron beam on phosphor particles. The appearance of flicker on CL phosphor screens changes with the irradiation time of the electron beam at the same current on the same phosphor screen. One does not perceive flicker images under irradiation of a scanning electron beam of 300 µA for 0.2 ns (4×10^5 electrons); and an appreciable flicker does appear under irradiation of the same electron beam for 2 ns (4×10^6 electrons). The threshold number of electrons changes with the conditions of phosphor screens. The details of the tangled conditions

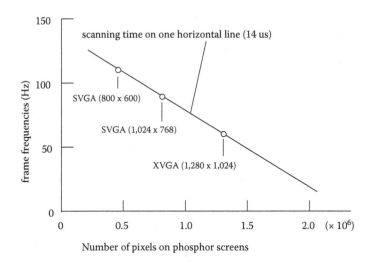

FIGURE 4.2
Minimum frame cycles for the perception of flicker of images on a 17-inch VTD screen as a function of information densities on the phosphor screen (resolution).

for generation of flickers on CL phosphor screens are discussed quantitatively in Section 4.4.

Regardless, VTD engineers empirically found the minimum frame frequencies for VTDs at different resolutions for practical flicker-free images, without any comprehension of the physical properties of the bulk of phosphor particles [106, 107]. Figure 4.2 shows a summary of the minimum frame frequencies that perceive flicker on the phosphor screen with different resolutions of a 17-inch VTD: SVGA (800 × 600), SVGA (1024 × 768), and XVGA (1280 × 1024).

As shown in Figure 4.2, the minimum frame frequencies for the perception of flicker is a constant scanning time for a horizontal line that is 14 μs for the 17-inch VTD. The constant scanning time of the horizontal line gives a constant irradiation time for a given phosphor particle in the CL screen. The time is calculable from the scanning speed of the electron beam, which is given by (horizontal length)/(scanning time of one horizontal time). Because the scanning speed is 2.4 × 10⁶ cm s⁻¹ for the 17-inch VTD, the irradiation time on a 4-μm particle is 0.2 ns. Under TV operating conditions (NTSC), the irradiation time on the same phosphor particle is 1 ns. The irradiation of an electron beam of 300 μA on the phosphor particle for 0.2 ns does not generate appreciable flicker, but the same phosphor particle exhibits flicker with irradiation for 1 ns. The results in Figure 4.2 indicate that the flicker of images on phosphor screens is related to the irradiation time and the magnitude of the electron beam on the phosphor particle in the screen; as well, the flicker is caused by neither the target accuracy of electron beam on phosphor pixel nor the fluctuation in terrestrial magnetism.

Additional confusion stems from the minimum frame frequencies in Figure 4.2, which change with the commercial phosphor powders used. The surface of commercial phosphor powders is deliberately contaminated with the microclusters (e.g., surface treatment and pigmentation).

The flicker (and moiré) of images on CL phosphor screens closely relates to the electric properties of the bulk of phosphor particles and microclusters on phosphor particles. The decay time of phosphors does not directly correlate with the generation of image flicker. A more likely possibility of moiré and flicker is the electron flow route at the phosphor screen. Before studying the electrical properties of phosphor screens, we must optimize the phosphor screen in an effort to remove further ambiguity regarding phosphor screens.

4.3 Powdered Screens Rather than Thin-Film Screens

Material science has developed from single crystals to thin films via sintered materials. Phosphor screens are also made by thin films. Because CL is generated in the bulk of phosphor crystals, the same mechanism is involved in thin-film screens and powdered phosphor screens. Contrary to expectations, it was empirically known before 1960 that powdered screens have luminances that are 5 times greater than those of thin-film screens. We may find a reason for the empirical results (i.e., 5 times) of powdered ZnS phosphor screens.

Phosphor screens are made by the arrangement of particles on the faceplate of CRTs in a defined area. Because the penetration depth ($\sim$0.5 μm) of the incident electron beam into the phosphor particles is shallower than the thickness of the thin film ($\sim$1μm) and phosphor particles (4 μm), the energy dumped onto the screens is the same for both screen types. One difference is the surface area irradiated by the incident electron beam. The surface area of the thin film is equal to that of the faceplate. The total surface area of particles exposed to electron irradiation is wider than the faceplate area. Calculations are given below.

We assume spherical particles. A spherical particle of diameter φ ($= d$) puts on a square (Figure 4.3). The surface area of the square S_0 is d^2. The surface area S of the spherical particle is

$$S = 4\pi(\varphi/2)^2 = \pi\varphi^2 = \pi S_0 \tag{4.2}$$

We can put four particles of diameter ($\varphi/2$) on the square (Figure 4.3B). The total surface area of the four particles $S_{1/2}$ is

$$S_{1/2} = 4 \times \pi(\varphi/2)^2 = \pi\varphi^2 = \pi S_0 \tag{4.3}$$

With particles of diameter d/n on the square, the total surface area $S_{1/n}$ is

$$S_{1/n} = n^2 \times \pi(\varphi/n)^2 = \pi\varphi^2 = \pi S_0 \tag{4.4}$$

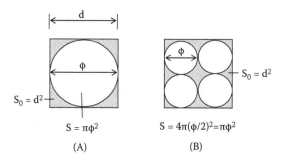

FIGURE 4.3
Total surface area S ($= S_0$) of particles arranged on a defined area $S_0 = d^2$. (A) is one particle and (B) is four particles.

Particles of the same diameter are densely arranged on the defined area in a single layer; the total surface area of the particles is given by πS_0, regardless of the diameter of the particle. Surface areas exposed to electrons from an electron gun are half that of the particles; 1.6 S_0 ($= 0.5\pi S_0$). There are gaps between particles arranged in a single layer, and the gaps do not emit CL. The projected gap area on the faceplate is 0.2 $\{= d^2 - \pi(\varphi/2)^2\}$ of S_0. The gaps should be covered by particles in a single layer for the optimization of screen luminance. Then, the total exposed surface area S_{total} is

$$S_{total} = 1.2 \times 1.6 \, S_0 = 1.9 \, S_0 \qquad (4.5)$$

The difference in total surface area (i.e., about twofold) does not explain the fivefold difference in the luminance. To explain this difference (i.e., 5 times), one should take into account the V_{eff} for EHs that migrate in the particles (4 μm). Then, the V_{eff} of the powdered phosphor screen is given by

$$V_{eff} = S_{total} \times 4 \ \mu m = 7.6 \, S_0 \qquad (4.6)$$

Considering that phosphor particles in the powder are not spherical particles of equal size, and that they distribute over a wide range of sizes and shapes, Equation (4.6) may explain the difference in screen luminance (i.e., 5 times) between powdered and thin-film phosphor screens. From the calculations it can be said that the brighter phosphor screens in CL devices are made from powdered phosphors, and that the optimal phosphor screen in terms of luminance is constructed with two layers of particles.

Figure 4.4(A) shows SEM pictures of a cross-section of the CL phosphor screen of commercial CRTs. The screen is constructed with thick layers. The preparation of a two-layer screen is difficult work. Figure 4.4(B) shows a cross-section of an experimental sample of the ideal phosphor screen, as an example. We must determine the flicker of images on the screen shown in Figure 4.4(A).

(A)   (B)

FIGURE 4.4
SEM pictures of cross-section of a commercial phosphor screen (A) and ideal screen in a high-resolution monochrome CRT (B).

4.4 Electric Properties of Bulk of CL Phosphor Particles in Screens

Notwithstanding, the electric properties of the bulk of CL phosphor particles and of phosphor screens are important items relating to the image quality on phosphor screens; and it has proven difficult to find reports relevant to this subject.

Figure 4.5 illustrates a simplified CRT structure and the electrical connection for CRT operation. The electron flows between cathodes and the

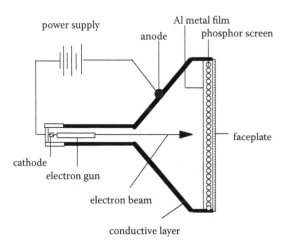

FIGURE 4.5
Simplified CRT structure for electric connection around the CRT.

aluminum (Al) metal film on the phosphor screen (inside of CRT), and between the anode and cathode (outside the CRT), are clear. The electron flow route at the phosphor screen is not yet clearly understood.

Two models have been considered for continuous electron flow at the phosphor screen. One is an electrical conductance through the phosphor particles, and the other is a collection of electrons from phosphor particles by Al metal film on phosphor screen. The models remain as assumptions.

First, we examine the electron flow at the phosphor screen by the electric conductance between the phosphor particles. The electric conductance α of a particle (crystal) is expressed by

$$\sigma = n_e \mu_e \ (\text{or } n_h \mu_h) \tag{4.7}$$

where n_e (and n_h) is the number of electrons (and holes) generated in the particle, and μ_e (and μ_h) is the mobility of the electron (and hole) in the particle, respectively. In phosphor particles, $\mu_e \gg \mu_h$; therefore, μ_e is considered for the electrical conductance. The value of μ_e in phosphor particles is very small, so that phosphor particles belong to the insulator. The phosphor screen is essentially constructed of insulator particles.

Under irradiation of a scanning electron beam on the phosphor particles, each entering electron generates a few thousand EHs in the particle; and with NTSC conditions, 4×10^5 electrons irradiate the 4-μm phosphor particle within a frame. With a large number of generated electrons (10^8) in the particle, the particle becomes a conductive particle (i.e., photoconductance).

In reality, the photoconductive particles in the screen are restricted in the phosphor particles arranged at the top layer of the screen in which incident electrons enter, and the entering electrons do not penetrate through the particles. Figure 4.6 schematically illustrates the phosphor layers on the substrate; the dark circles express the photoconductive particles, and the white circles are identify the insulators. If the electric conductance is considered between the particles at the top layer, further experiments can be performed to determine the electric conductance between the conductive particles [21].

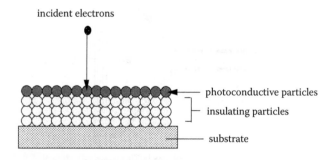

incident electrons

photoconductive particles

insulating particles

substrate

FIGURE 4.6

The structure of a phosphor screen, explaining the photoconductive particles at the top layer (black circles) and the insulating particles between the top layer and the substrate (open circles).

The phosphor particles are carefully arranged on a glass substrate (insulator). These particles, however, do not generate any electrical conductance in the screen, although each particle possesses high conductance. This can be examined using a highly conductive SnO_2:Sb powder (average particle size is 2 μm). A screen (0.1-mm thickness) of the SnO_2:Sb powder is made on a glass substrate by sedimentation, without any binder, as with phosphor screens. A very low conductance (1/500 kΩ) is measured with the screen. If the screen is pressed with a finger, the electrical conductance of the screen increases. The screen shows good conduction (1/10Ω) under 100 psi [21]. Thus, the carefully arranged particles in the screen do not have any electrical conductance. Furthermore, the surface of the commercial phosphor particles are covered with a nonluminescent layer that adheres to microclusters of SiO_2 and pigments. Phosphor particles in the screen have gaps filled by insulators.

Then we can discuss the direct collection of electrons from the photoconductive particles at the top layer by Al metal film on the phosphor screen. To collect the electrons from a conductive particle, the Al film must have ohmic contact with the particles. It is well known in the study of compound semiconductors (single crystals) that Al metal does not form ohmic contact with II–VI compounds. The ohmic contact with II–VI crystals is only observed with melted Ga heated above 400°C. In addition to the ohmic contact, there are gaps between the Al metal film and phosphor particles. Furthermore, both sides of the Al film are covered with an Al_2O_3 layer (alumina, white body color), which is a typical insulator, as a consequence of the oxidization of the Al film heated at 430°C in air in the CRT production process. Then one can say that the Al film on the phosphor screen cannot collect the electrons from the particles at the top layer of the screen. This statement is supported by evidence that the sample for SEM observation is covered with Au film, and that the sample with Al film cannot be used for SEM observation.

The above evidence may be sufficient for the unverified models. We must now find out an electron flow route at the phosphor screen in CRTs in order to close the electric circuit. An overlooked phenomenon is the presence of secondary electrons. Phosphor particles inevitably emit secondary electrons in vacuum. Figure 4.7 illustrates a model of the electron flow route at the phosphor screen via the collection of secondary electrons in a vacuum [21].

Many reports refer the ratio of the secondary electron emission to incident electrons (i.e., the δ ratio), which has two cross-voltages at $\delta = 1.0$: low and high anode voltages [109]. The δ model derived from measurements of the CL of Zn_2SiO_4:Mn phosphors some 60 years ago. The δ model must be revised as a result of more current knowledge.

When incident electrons enter a crystal, the penetrating electrons lose their energy via elastic and inelastic collisions with the lattice ions along the electron trajectories, generating Auger electrons and secondary electrons of energy less than that of the primary electrons. These secondary electrons generate other secondary electrons of energy less than that of the previous

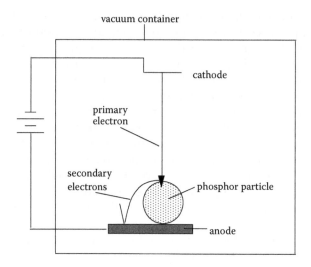

FIGURE 4.7
Model of electric flow route at phosphor screen by a collection of secondary electrons emitted from phosphor particles in a CL vacuum device.

secondary electrons — that is, applying the cascade model of the energy depletion process. The internally generated secondary electrons form the electron gas plasma. The mean free path of the plasma electrons in the crystal is about 10 Å [110]. Hence, only the secondary electrons generated in the surface volume shallower than 10 Å depth will escape from the crystal surface. The spectrum of the detected secondary electrons in front of the crystal is independent of the energy of the incident electrons. A large part of the detected secondary electrons are widely distributed at energies below 50 eV, and the peak lies at around a few electron-volts (eV). These electrons are called *true secondary electrons* and are used in the scanning electron microscope. Each electron entering the crystal must emit multiple secondary electrons ($\delta > 1.0$). There is no cross-voltage ($\delta > 1.0$). The classical textbook takes the threshold voltage for the appearance of luminescence, V_{th} of Zn_2SiO_4:Mn phosphor (around 110 V), assuming that the incident electrons have penetrated the phosphor particles. As described in Section 2.7, the penetrating electrons do not generate CL below V_{th}.

The emission of true secondary electrons leaves a corresponding number of holes in the surface volume of the crystal. The positive field of the holes inside the crystal extends outside the crystal and attracts the true secondary electrons. If the true secondary electrons do not re-enter the crystal, the electrons can bind outside at a short distance from the crystal surface. The bound electrons are the surface-bound-electrons (SBEs) [111]. Eventually, a negatively charged electron cloud (space charge) forms in front of the insulator after the incident electrons have entered. The trajectory of incoming electrons on the insulators is disturbed by the negative field of the SBEs.

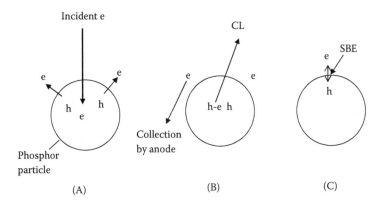

FIGURE 4.8
Model for the formation of SBEs on the front of the surface of a phosphor particle: one entering electron emits two secondary electrons from the phosphor particle, leaving two holes in the particle (A). The anode collects one electron at the outside of the particle, and one electron recombines with a hole at the luminescence center (B). Then, one electron in the front of the particle binds with a hole in the surface volume of particle, forming the SBE (C).

Phosphor particles are insulators, but phosphors are a particular insulator type that contains the recombination centers of the EHs.

Figure 4.8 illustrates a model of a phosphor particle under electron irra-diation. We assume that one entering electron emits two secondary electrons [113], leaving two holes in the surface volume of the particle (Figure 4.8A). One hole radiatively (or nonradiatively) recombines with the entering elec-tron at a recombination center. One secondary electron outside the particle is collected by the anode underneath the phosphor screen to close the elec-tron-flow circuit at the phosphor screen (Figure 4.8B). The particle holds one hole in the surface volume and one electron outside the particle — that is, the SBE (Figure 4.8C). The number of SBEs on a phosphor particle signifi-cantly decreases by recombination of EHs as compared with the SBEs on an insulator, but some number of SBEs remain on the surface of the phosphor particle.

We can calculate the electric field (E_h) generated by a hole in the surface volume of the phosphor particle. The electric field E_h at a distance r from the hole is expressed by

$$E_h = Q_h \cdot (4\pi \, \varepsilon_0 \, r^3)^{-1} \tag{4.8}$$

where Q_h is the charge of a hole (1.60×10^{-19} coulomb) and ε_0 is the dielectric constant in vacuum (8.85×10^{-12}). Because the penetration depth d_p of inci-dent electrons is negligibly small as compared with particle size φ (e.g., $d_p << \varphi$), we can use the distance r from phosphor particles, instead of the accurate distance from the holes. Using a microprobe, an electron cloud was detected at around 5 μm (5×10^{-4} cm) from the surface of the phosphor particles. Assuming that the secondary electron can remain at 5 μm (5×10^{-4} cm) from the surface of particles, E_h is estimated as 11 V cm^{-1}.

In practice, many holes are generated in the surface volume of the particles. With N holes, the electric filed ΣE_h is

$$\Sigma E_h = \Sigma Q_h \cdot (4\pi \, \varepsilon_0 \, r^3)^{-1} \qquad (4.9)$$

We can calculate ΣE_h from the dependence curves of screen luminance on electron beams (DSLEB curves). The measurement of DSLEB curves is made using a 0.5-inch miniature CRT (1-cm^2 screen size) with VGA conditions [21]. The spot size of the electron beam is 17 μm. The phosphor powder has a clean surface. Figure 4.9 shows the screen luminance of Y_2O_2S:Tb:Eu ($\varphi = 2$ μm) phosphor screens of two- and six-particle layers on a glass substrate, under irradiation of various electron beam currents (0.3 to 10 μA·(spot)$^{-1}$). The screen luminance of the two-layer screen was 30% higher than that of the six-layer screen. This is because only phosphor particles at the top layer of the screen (gun side) emit CL, and there is optical loss by scattering in the six-layer screen — about 8% loss per layer ($0.32 = 4 \times 0.08$) [5]. Our attention focuses now on (1) the inflection points and (2) slopes of the curves, rather than the screen luminance. In CRTs, the anode is situated on the side wall of the CRT envelope, and phosphor powder screens on the glass faceplate. With this screen configuration, the DSLEB curves of two- and six-layer screens inflect at 4 μA·(spot)$^{-1}$, which is independent of the anode voltage (2 to 10 kV) at the side wall. Below 4 μA · (spot)$^{-1}$, the slope is 1.0; above 4 μA · (spot)$^{-1}$, the slope is 0.5. We can calculate ΣE_h at the inflected electron beam current, 4 μA, of the DSLEB curve.

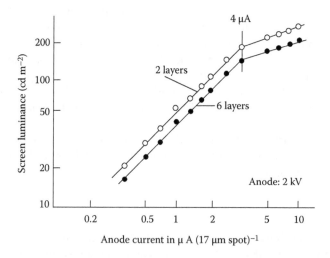

FIGURE 4.9
DSLEB curves of Y_2O_2S:Tb:Eu phosphor screens of different thickness (two and six layers of particles) in a 0.5-inch CRT, as a function of the irradiated densities of an electron beam of energy 2 kV.

An electron beam of 4 μA contains 2.5×10^{13} electrons (sec and spot)$^{-1}$. The irradiation duration of the 17-μm spot on the phosphor particle is 20 ns. Therefore, the number of the irradiated electrons of the 17-μm spot on the phosphor particle for 20 ns is 5×10^5 electrons ($= 2.5 \times 10^{13} \times 20 \times 10^{-9}$). Then, we have

$$\Sigma E_h = 11 \times 5 \times 10^5 = 5 \times 10^6 \ (\text{V cm}^{-1}) \tag{4.10}$$

We assume that the negative field of SBEs is equal to the ΣE_h. As the anode field E_a is greater than ΣE_h, the negative field of the SBEs is concealed by E_a. Incoming incident electrons enter the phosphor particle without disturbing the trajectory, giving rise to flicker-free images on the phosphor screen. In the case of Figure 4.9, the anode is at the side wall, and E_a applies only to the top-layer phosphor particles, independent of screen thickness. This is the reason that DSLEB curves inflect at 4 μA for the screens of two and six layers. We cannot calculate the E_a of anode configurations at the side wall of a CRT.

When the anode resides underneath the phosphor screen, E_a is given by

$$E_a = V_a \, r_a^{-1} \tag{4.11}$$

where r_a is the distance from the anode. The E_a at the top layer of the phosphor screen can be expressed as the number of layers N_L of particles multiplied by the average particle size ($r_a = N_L \, \varphi$), instead of r_a [5]. Therefore,

$$E_a = V_a \cdot (N_L \varphi)^{-1} \tag{4.12}$$

Figure 4.10 shows the DSLEB curves of Y_2O_2S:Eu phosphor screens ($\varphi = 2 \ \mu$m) with an anode potential of 7 kV. The DSLEB curve of the six-layer

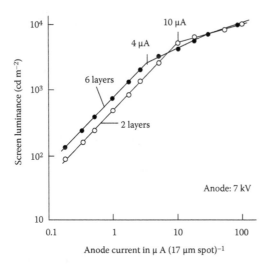

FIGURE 4.10
DSLEB curves of Y_2O_2S:Tb:Eu phosphor screens having two and six layers of particles in a 0.5-inch CRT, as a function of the irradiated densities of an electron beam of energy 7 kV.

screen inflects at $J = 4$ μA·cm^{-2} ($\Sigma E_h = 5 \times 10^6$ V cm^{-1}). The calculated E_a of the six-layer screen is 6×10^6 V cm^{-1} {$= 7 \times 10^3 \times (6 \times 2 \times 10^{-4})^{-1}$}, giving $\Sigma E_h = E_a$ at $J = 4$ μA cm^{-2}. The E_a of the two-layer screen is 1.7×10^7 (V cm^{-1}). The DSLEB curve of the two-layer screen inflects at 10 μA cm^{-2}, giving $\Sigma E_h = 1.2 \times 10^7$ (V cm^{-1}). Thus, the calculated ΣE_h agrees with E_a on the particles at the top layer of the phosphor screen. Below the inflection points, the phosphor screen displays flicker-free images. The results positively indicate that the flicker-free images are only obtained with a thin phosphor screen (possibly two layers).

If the particles do not contain the recombination centers of EHs (e.g., insulator), the holes in the surface volume of the insulator (corresponding to ΣE_h) accumulate with increasing irradiation time of the electron beam on the insulator. A typical insulator is SiO_2. When the SiO_2 microclusters adhere to the surface of the phosphor particles, the SBEs effectively shield the phosphor particle negatively. To make a solid confirmation of the shield of individual phosphor particles by SBEs on insulators, SiO_2 microclusters adhere to the surface of the green-emitting Y_2O_2S:Tb phosphor powder (which has a clean surface), and the red-emitting Y_2O_2S:Eu phosphor powder, which does not have SiO_2 microclusters on the surface. The Y_2O_2S:Eu phosphor emits red CL with V_a greater than 110 V. Both phosphors have the same average particle sizes. The phosphors are mechanically mixed in a 1:1 ratio by weight. The mixed phosphor powder was screened on the faceplate of a CRT with two layers of particles. The screen emits the red CL, as V_a is less than 1000 V. The phosphor screen emits yellow CL with V_a greater than 3000 V, indicating that the both red and green phosphor particles emit via the incident electrons. By observing the yellow CL screen under an optical microscope ($> \times 100$), the green CL spots were flicker, whereas no flicker was observed for the red CL spots. Hence, it is thought that the flickering images on CL phosphor screens is caused by the disturbance of the trajectory of incoming electrons by the negative charge of the SBEs on phosphor particles. The results definitely indicate that a flicker-free CL phosphor screen is only obtained with phosphor particles having a clean surface.

In practical CRTs, the faceplate is not conductive. Color CRTs have a black matrix (BM) of conductive carbon powder of 0.2-μm thickness on the faceplate. The BM may work as an anode for a two-layer screen. Although the phosphor screens are made with pigmented phosphor powders, the flicker of images is significantly suppressed to an acceptable level with two-layer screens on the BM if the screen luminance is less than 150 cd m^{-2}. The screen exhibits flickering images when the screen luminance is above 200 cd m^{-2}.

4.5 Optical Properties of Bulk Phosphor Particles

Despite the fact that the image quality of phosphor screens at high luminance is determined by optical scattering on the surface of the phosphor particles, optical scattering by phosphor particles is commonly overlooked in the study of phosphor screens in CL devices. Optical scattering is governed by the physical properties of bulk host crystals, which are asymmetric. As described in Section 2.2, and shown in Table 2.2, the asymmetric crystals have a large dielectric constant ε that relates to the index of refraction n (= ε^2). The large n values are not only a benefit that gives a wide-view angle of images on CL phosphor screens, but also lead to the disadvantages of smear, low contrast ratio, and poor color fidelity of the images. The theoretical details of the optical scattering of emitted CL light were discussed in Section 2.2. Emitted CL light reaches viewers after reflection on the surface of phosphor particles arranged between the top layer and the substrate. Now we discuss some practical phosphor screens.

Color phosphor screens are constructed with patterned screens, which are separated by the BM. Because the BM is a thin layer (<0.5 μm), the phosphor particles (φ = 4 μm) are screened on the BM, and the BM does not separate the phosphor screens optically. Figure 4.11 shows a photo (in reflection mode) of color phosphor screens. The gaps between patterned screens lengthen the mean free path of the scattering light. Aluminum metal film on phosphor screens further lengthens the mean free path. Consequently, the color phosphor screen receives the scattered CL light from neighboring phosphor screen pixels in different colors, as illustrated in Figure 4.12. When the phosphor screen receives scattered CL light from neighboring screens in different color pixels, images on the phosphor screen smear and colored image whiten because of the mixing of CL colors.

In color CRTs, the images on phosphor screens are evaluated with a balanced color, which is given by the theory of color mixing. Figure 4.13 gives the color coordinates (dark circles) of ZnS:Ag:Cl blue, ZnS:Cu:Al green, and

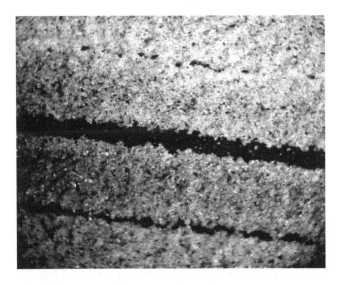

FIGURE 4.11
Patterned color phosphor screens on the faceplate that has a black matrix of 0.5 μm.

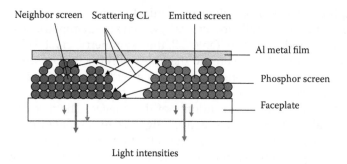

Neighbor screen   Scattering CL   Emitted screen

Al metal film

Phosphor screen

Faceplate

Light intensities

FIGURE 4.12
Explanation of smeared images by scattered CL light from neighboring phosphor screen pixels in different colors. Color contamination is caused by scattered light that has a long mean free path between patterned phosphor screens and Al metal film on the phosphor screen.

Y_2O_2S:Eu red phosphors on an x–y color diagram. Balanced color images are given by equal (l $Y \cdot y^{-1}$) for each color, where l is the distance between white (x = 0.33, y = 0.33) and each color (l_r, l_g, and l_b) on an x–y color diagram, and ($Y \cdot y^{-1}$) is the lumen weight. Y is the luminance and y is the color coordinate on the x–y color diagram. Because $l_r \approx l_g \approx l_b$, and the balanced color is given by the lumen weights (Y y^{-1}) of each color: (Y_b y_b^{-1}) ≈ (Y_g y_g^{-1}) ≈ (Y_r y_r^{-1}). The y-values are y_g = 0.60, y_r = 0.34, and y_b = 0.07, respectively. For equal (Y y^{-1}), a large Y value is required for the green screen and a small Y

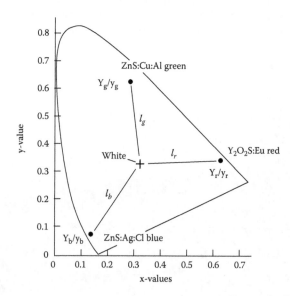

FIGURE 4.13
Color coordinates of blue ZnS:Ag:Cl, green ZnS:Cu:Cl, and red Y_2O_2S:Eu phosphors on the x–y color diagram, and an explanation of the importance of lumen weight (Y y^{-1}) in the color balance of color phosphor screen that has $L_B = L_G = L_R$.

value (1/12 of green) is required for the blue screen. The luminance of the balanced white color on a phosphor screen is roughly composed of 62% green light, 30% red light, and 8% blue light. As a consequence, the blue and red phosphor screen pixels pose a serious problem in terms of the scattered CL from green screen pixels, and the green phosphor pixels are whitened by small amounts of scattered CL from the blue and red screen pixels.

If the colored microclusters adhere to the surface of phosphor particles (i.e., pigmentation), the color pigment can absorb some amount of scattered CL light from neighboring screens. For example, the Fe_2O_3 red pigment on red phosphor particles absorbs some amount of scattered green and blue light, so that the scattered CL light markedly decreases from the red phosphor pixels. Similarly, $CaAl_2O_4$ blue pigment on blue phosphor particles absorbs scattered green and red light, giving the blue CL light. The color fidelity of image pixels is markedly improved by the pigmentation of red and blue phosphor powder. There is no availability for green pigment. The green screen has a whitening problem due to the scattered blue and red CL light from neighbors. The purpose of the pigmentation of phosphor particles is absorption of the scattered CL light from neighboring phosphor screens, and it is limited to screen luminances below 200 cd m^{-2}. It is not the absorption of ambient light in a room, which was previously considered [114]. The faceplate of the CRT is well tinted for absorption of ambient light.

As described in Section 4.4, the adhesion of the pigments (insulators) to the surface of particles gives rise to the problem of flicker images. As a solution to this pigmentation, the insertion of color filter layers between phosphor screens and the faceplate has been applied to color phosphor screen [113]. In this case, the phosphor particles may have a clean surface. The optical effect of the insertion of a color filter layer is the same with the pigmentation.

A perfect solution to the scattering of emitted CL light to neighboring phosphor screen pixels can be achieved by surrounding the patterned phosphor screen pixels with a thick BM, that is, optical isolation of individual patterned phosphor screen pixels with a thick BM [24]. Figure 4.14 illustrates

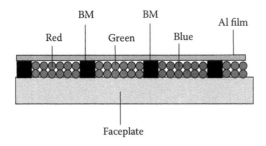

FIGURE 4.14
Illustration of the screen structure that confines emitted CL light in patterned screen pixels by the light-absorbing BM layer.

the structure of patterned phosphor screen pixels that are surrounded by the thick BM. The BM not only confines the emitted CL light in patterned screens, but also works as an anode that effectively conceals the negative field by SBEs on phosphor screens. Because phosphor powders have a clean surface, the image quality on the improved color phosphor screens is equal to the printed images on sheets of paper, even for a high-light scene.

5

The Screening of Phosphor Powders on Faceplates

Powdered phosphors are used as the screen that arranges phosphor particles (φ = 4 μm on average) with thicknesses of approximately 4 mg cm^{-2} on the faceplates of display devices. Phosphor screens in cathode ray tubes (CRTs) have many screening problems related to the phosphor powders. Here we delve into the screening problems of phosphor powders on the faceplates of color CRTs. The results are applicable to phosphor screens of photoluminescent (PL) devices.

One gram of phosphor powder contains an average of 7×10^9 particles [= $(\pi\varphi^3\rho/6)^{-1}$, where ρ = 4.1 for ZnS]. The area of the faceplate of a 25-inch CRT is 1850 cm^2; therefore, 5×10^{10} particles are arranged on the 25-inch faceplate. One cannot arrange 5×10^{10} individual particles on a faceplate. Discussions of the screening of phosphor powders are made by statistical averages of a huge number of cathodoluminescent (CL) phosphor particles on the faceplate. The screening results of phosphor powders represent the average trials of 10^{10} particles. Color phosphor screens on faceplates are sequentially produced by photolithography applied to dried color phosphor screens. Accordingly, we separately discuss (1) the screening of phosphor powder on the substrate, and (2) the patterning of dried phosphor screen by photolithography.

Because as many as 200 million CRTs are produced annually on a worldwide basis, it was believed that the screening technologies of CL phosphor powders on the faceplates of CRTs are well optimized scientifically — that is, state-of-the-art. In reality, there are only a few publications on the screening mechanisms and patterning of dried phosphor screens [114–120], and the screening of phosphor powder on faceplates remains somewhat of an ambiguity. One difficulty encountered in the scientific study arises from the fact that there is as yet no ideal phosphor screen. We may find out the reason for the difficulty in finding the ideal phosphor screen. A first question that arises in the screening process of color CRT production concerns the aging process of PVA (polyvinyl alcohol) phosphor slurries for a long period of time, sometimes all day long. The reason for this is the residuals in the commercial phosphor powders. In addition to residuals, the phosphor particles of commercial phosphor powders are widely distributed with respect to size and shape. Chapter 5 delves into the screening problems associated with the use of commercial phosphor powders.

5.1   The Ideal Phosphor Screen

When phosphor screens are produced by established screening technology, the obtained screens are always of uneven thickness. Figure 5.1 shows a microscopic picture (700X) of color phosphor screens on the faceplate of a 20-inch CRT. The screen exhibits irregularity in thickness, which makes it difficult to study scientifically. For scientific study, the phosphor screen should be even in terms of thickness, that is, an optically continuous medium.

If phosphor particles have a clean surface chemically, phosphor particles will disperse in deionized (DI) water with the addition of a very small amount of a disperser (surfactant), such as polyvinyl alcohol (PVA) resin, disodium hydrogen phosphate (Na_2HPO_4), potassium silicate (xK_2O-SiO_2), etc. For example, the addition of 0.5 ml of 5% PVA solution to 100 ml DI water results in a perfect dispersion of phosphor particles in stirred water. The role of the disperser only separates the particles in water; it does not bind particles on the substrate, such as binding via an Si–O chain. The suspension, in which the particles are perfectly dispersed, is poured into a container that has a substrate at the bottom. Suspended particles settle to the bottom in water, according to Stokes' law. As illustrated in Figure 5.2, the settled particles on the substrate form a phosphor screen of even thickness. Screen thickness is controlled by the amount of suspended phosphor powder. Special attention is required for adhesion of the settled particles onto the substrate.

Just-settled particles on the substrate exhibit weak adhesion with water in the gaps between the particles. For strong adhesion of particles to the substrate, the settled screen leaves for a period of time longer than 60 minutes.

FIGURE 5.1
Microscope image (700X) (in reflection mode) of currently produced color phosphor screens in CRTs.

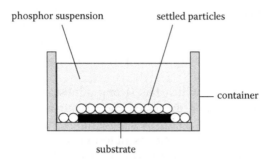

FIGURE 5.2
Schematic illustration of settled particles on the substrate in a container, wherein phosphor particles perfectly disperse throughout the suspension.

Excess water in the gaps slowly moves out and small particles move into the gaps, that is, wedging of the settled particles. Wedged particles do not move in position when the top, clear water is slowly removed from the container. The wet screen is placed in a heated oven at 50°C and then the screen is heated slowly to 90°C until dry. Figure 5.3 shows the SEM

FIGURE 5.3
SEM images of a cross-section and an overview of the ideal phosphor screen.

picture (2000X) of a cross-section and an overview of a nearly ideal phosphor screen of $Y_2O_2S:Tb:Eu$, which has a clean surface. Similar results are also obtained with adhesive substrates [119]. One difficulty encountered in the use of adhesive substrates is the decomposition of large amounts of adhesive organic material by heat.

The phosphor powder, which has strong adhesion of particles onto the substrate, can be preliminarily tested by measuring a settled density of suspended particles (5 g) in a narrow testing tube (5-mm inner diameter and 30-cm length). The settled density is determined by measuring the well-settled particles in the testing tube. The results of the settled density correspond with the adhesion of the wedged particles onto the substrate. However, the results do not correlate with the adhesion of just-settled particles onto the substrate and rapidly dried screen, as in ordinary CRT production.

5.2   Practical Problems in Drying the Unwedged Phosphor Screen

Because the container is used for the settlement of the phosphor screen, we cannot make a microscopic study of the adhesion of the just-settled particles on the substrate. Fortunately, a thin phosphor screen (~ one layer) is made on a small substrate (12 cm²) by a mesh screen technique, using a mixture of phosphor powder, water, and disperser. The screen slowly dries via the evaporation of water. Figure 5.4 shows micrographs of dried phosphor screens of wedged (Figure 5.4A) and unwedged (Figure 5.4B) phosphor particles on a glass substrate.

Observation of a drying screen under a microscope (>200X) provides the following information.. The unwedged screen slowly dries by evaporation of water. In the final process of drying the screen, the water suddenly and quickly moves on the substrate. If the particles are not settled on the substrate (unwedged particles), moving water carries floating particles. The carried particles pile up around the large particles (anchor particles) that adhere to the substrate. Moving water on the substrate inevitably generates piled-up particles (e.g., clumped particles) and pinholes in the unwedged phosphor screen. The movement of water on the wet screen speeds up with increasing temperature.

When the dried screen is observed under the transmission of light, many tiny dark spots are detected on the screen. The dark spots were misidentified as clumped (aggregated) particles in the PVA phosphor slurry. Then, the dispersion of phosphor particles in the PVA slurry was considered the major screening problem. The piled-up particles cannot be eliminated from the dried screen by the addition of dispersers. The cause is piled-up particles of the separated particles during the drying process.

In practice, a further complication stems from the microclusters adhered to the surface of phosphor particles (surface treatment). If the surface of the particles adheres to the microclusters, as in commercial CL phosphors, then

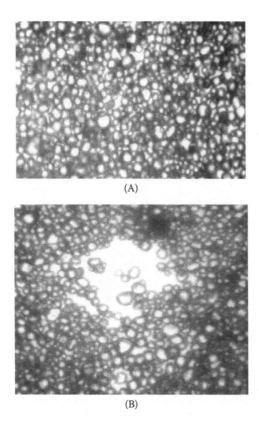

(A)

(B)

FIGURE 5.4
Micrograph of dried PVA phosphor screens on glass substrate with a single layer: (A) a well-wedged phosphor screen, and (B) an unwedged phosphor screen that has dried rapidly.

the microclusters determine the contact gaps between particles, which hold water by capillarity. The particles with microclusters are not wedged between the settled particles on the substrate. In drying the screen, the moving water suddenly and strongly sucks out the water from the gaps against the strength of capillary action, and the water carries away the floating particles. This is an important observation in the screening process. The phosphor particles with microclusters (surface treatment) are inadequate phosphor particles for the preparation of an ideal screen by settlement in the container.

5.3   Optimal Thickness of Phosphor Screens in CL Devices

Now we know how to make the ideal phosphor screen using wedged particles. Using phosphor screens of even thickness, we can find the optimal screen thickness that gives maximum CL output. There are two kinds of

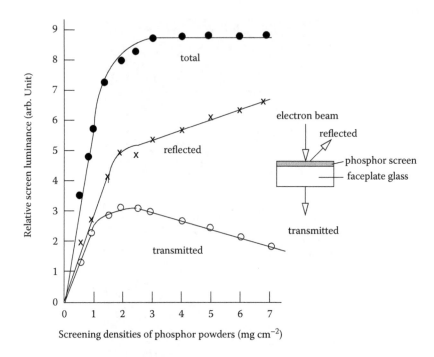

FIGURE 5.5
Transmitted, reflected, and total screen luminance of CL phosphor screens as a function of the screening density of the phosphor powder (mg cm^{-2}) on the substrate. Screens are made by sedimentation without K_2O-SiO_2 disperser, and the settled screens leave for 90 min. for wedging.

screen luminance: (1) the screen luminance measured at the faceplate (i.e., transmission, T-mode), and (2) the screen luminance measured at the electron-beam side (i.e., reflection, R-mode). The total screen luminance emitted in phosphor screens is given by addition of these two luminances. Figure 5.5 shows the curves of screen luminance of T-mode, R-mode, and total screen luminance emitted from a phosphor screen, as a function of the screen density (mg cm^{-2}) of phosphor powders. The average particle size is 3.5 μm. Practically, images on phosphor screens are viewed in T-mode, which gives the optimal screening density, $w_{opt} = 2$ mg cm^{-2}, in Figure 5.5.

The values of w_{opt} in T-mode luminance markedly change with the φ of phosphor powders. Figure 5.6A shows the curves of T-mode luminance with screening densities of phosphors of various φ (= 3, 5, 7, and 12 μm). The optical properties of phosphor screens are determined by the total surface area of the phosphor particles arranged in a defined area [5]. As described in Section 4.3, it is S_0 per layer of particles and it is constant for a defined faceplate. Therefore, the T-mode screen luminance should be based on the number of particles N_L, instead of the screening weight density. Figure 5.6(B) is the re-plotted curve of Figure 5.6(A), showing that all the data in Figure 5.6(A) fit on one curve that is a function of N_L. The optimal screen thickness is

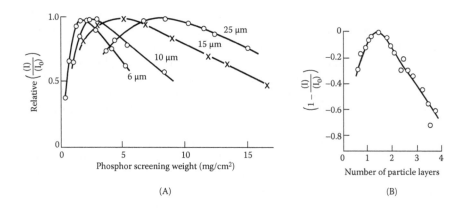

FIGURE 5.6
Relative screen luminance (transmission mode) of phosphor screens by different particle sizes (3, 5, 7, and 12 μm) as (A) a function of the screening densities of phosphor powder, and (B) a function of number of layers of particles N_L.

$N_L = 1.5$, and is independent of φ. Note that the results in Figure 5.6 are only obtained with ideal phosphor screens, and $N_L = 1.5$ is the average number calculated from the screened weight of the powder. If the phosphor screens are produced with the screening facilities of CRT production, the results differ from the results in Figure 5.6; that is, a high N_L value.

5.4 Particle Size for High Resolution of Monochrome CL Images

The next concern is image resolution, that is, the minimum size of the CL spot on the ideal phosphor screen. Calculating appropriate particle sizes for high resolution is performed with images on monochrome CL phosphor screens. If incident electrons of diameter φ_e irradiate onto the phosphor screen, the diameter of the emitted area φ_{em} on the phosphor screen is wider than $\varphi_e(\varphi_e < \varphi_{em})$. From a geometric analysis of φ_{em} on the phosphor screen, the widening is limited to one phosphor particle φ, rather than two particles [118]. Hence, φ_{em} is given by

$$\varphi_{em} = \varphi_e + \varphi = \varphi_e \left(1 + \varphi/\varphi_e\right) \tag{5.1}$$

If φ/φ_e is smaller than 0.1, the resolution of CL images is accepted. In ordinary CRTs, φ_e is usually 200 μm, and $\varphi = 4$ μm, so that the emitted size φ_{em} on phosphor screens is solely determined by φ_e.

Only phosphor particles that are arranged on the first layer of the phosphor screen emit CL, with a shallow penetration depth (0.5 μm) of incident electrons (0.5 μm << 4 μm). Emitted CL light should travel through the phosphor screen on the faceplate glass to reach the viewer. The emitted CL light entering the faceplate glass has internal reflection (about 7%). With tinted glass, the internal reflection decreases to a negligible level in the discussion of image resolution. Therefore, the image size is primarily determined by the optical properties of the phosphor screen. The phosphor particle has a large index of refraction (n ≈ 2.4). The traveling CL light reflects on the surface of phosphor particles and the residuals penetrate into the phosphor particles. The amount of reflected CL light (R) is 40% [R = (n − 1)(n + 1)$^{-1}$ = 0.4], and 60% of the CL light penetrates the phosphor particle. The emitted CL light in the particle and the entering CL light from other particles into the particle exit the particle after experiencing internal multi-reflections in the particle (lighting particle). The lighting particles and traveling CL light in the phosphor screen give rise to a wide viewing angle for CL images. To simplify the calculation of the scattered CL light in the phosphor screen, one can consider the following model for the calculation: an emitting particle in the top layer makes contact with four nonemitting particles in the second layer, as illustrated in Figure 5.7.

The width of the scattered light (WSL) of the emitted CL from a single particle is given by

$$WSL = 2\varphi + g \tag{5.2}$$

where g is the packing gap between particles. With the screen comprising N_L particle layers, the WSL for a single emitting particle will be

$$WSL = N_L (2\varphi + g) \tag{5.3}$$

The total width of the observed spot (W_{ob}) is given by

$$W_{ob} = \varphi_{em} + WSL = \varphi_{em} + N_L (2\varphi + g) \tag{5.4}$$

For the minimization of W_{ob}, (1) the use of a small φ_{em}, (2) the minimization of N_L, (3) the elimination of g, and (4) a small φ have been considered. In

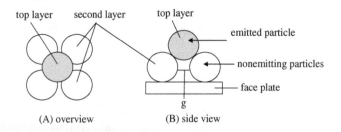

top layer   second layer   top layer

                                              emitted particle

                                              nonemitting particles

                                              face plate

                                  g

(A) overview   (B) side view

FIGURE 5.7
Schematic illustration of the scattering of emitted CL light from a particle at the top layer by four particles in the second layer.

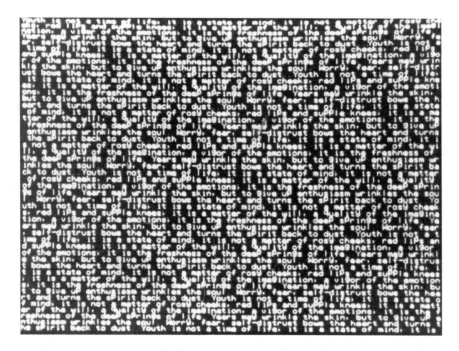

FIGURE 5.8
Photo (16X) of VGA images on a 1-cm² screen in a miniature monochrome CRT; 307,200 pixels cm⁻² [118].

the wedged phosphor screen, the gap is nearly zero (g 0). The preparation of $N_L = 1.5$ is very difficult; a practical phosphor screen can be prepared with $N_L = 2$, by sedimentation of dispersed particles in water. Then, $2 N_L \varphi = 16 \mu$m, less than 10% of φ_e (200 μm). The minimum light spot of ordinary monochrome CRTs is predominantly determined by φ_e. If the screen is made of seven layers, then $2N_L \varphi = 56 \mu$m. The particle size φ influences W_{ob}.

If the phosphor screen is made with $\varphi = 2 \mu$m and $N_L = 2$, then $2 N_L \varphi = 8 \mu$m. We can obtain $W_{ob} = 18 \mu$m with $\varphi_e = 10 \mu$m, which is an extremely high resolution for a monochrome CRT. The narrow electron beam size is obtained by focusing two magnetic rings, instead of the electrostatic focus electrodes used in ordinary CRTs. Figure 5.8 shows a photo (16X) of the mono-chrome images (black and white) on a phosphor screen in 1 cm² that gives an image density of 307,200 pixels cm⁻² [23]. Some letters in Figure 5.8 are smeared due to the resolution of the photo-film, not the images on the phosphor screen. This is the highest image density ever on a monochrome CRT screen, with an ideal phosphor screen. As a reference, the highest image density of XVGA (1280 × 1024 pixels) on a 20-inch screen is only 1092 pixels cm⁻².

The resolution of patterned sizes of color phosphor screens is basically the same as that of monochrome phosphor screens, by substituting particle sizes of the monochrome screen for dot sizes (or stripe widths) of color phosphor screens.

5.5   Photolithography on Ideal Phosphor Screens

Now we discuss the patterning of phosphor screens by photolithography on a dried PVA phosphor screen that is photosensitized by ammonium dichromate (ADC). Patterned color phosphor screens are sequentially made on faceplates. In a dried PVA phosphor screen, each phosphor particle is covered with PVA film (0.7 to 1 μm thick). Photolithography occurs by absorption of incident UV light by the dried PVA film on the phosphor particles.

The ADC in solution dehydrates into three segments: (1) $Cr_2O_7^{2-}$, (2) CrO_4^{2-}, and (3) $HCrO_4^-$, depending on the pH value. Only $HCrO_4^-$, which is present in the pH range 4.5 < pH < 8.0, photosensitizes PVA resin. The concentrations of $HCrO_4^-$ in solution sharply change (10 times) with small changes in pH between 5.5 and 6.5, although added ADC concentrations are rigidly controlled (±1%). The pH value in the PVA-phosphor slurry should be carefully adjusted to around 7.0 to achieve dispersion of the phosphor particles. The PVA film in the dried phosphor screen (pH = 7.0) has a low photosensitivity.

The photoreaction of PVA resin with ADC is well known in the area of photochemistry. As shown in Figure 5.9, the photosensitized PVA film, which has absorbed the UV light, becomes transparent to the incident UV light. Because the photoreaction of $HCrO_4^-$ is an irreversible process, Beer-Lambert's law is not applicable to the absorption of incident light. The amount of photoreacted PVA film (P_{react}) accumulates with the absorption of UV light, for which Bunsen-Roscoe's law is applicable:

$$P_{reac} = kIt \qquad (5.5)$$

where I is the intensity of the incident UV light and t is the exposure time, so that (It) is the exposure dose. Incident UV light that penetrates the phosphor

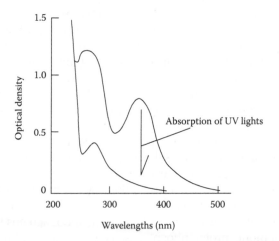

FIGURE 5.9
Absorption spectra of photosensitized PVA film before and after exposure to UV irradiation.

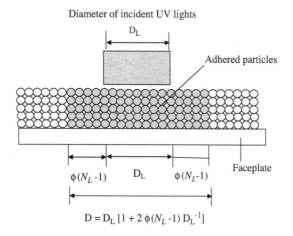

FIGURE 5.10
Schematic illustration of the pattern size D of a PVA phosphor screen in N_L layers, which adhere to the substrate by photolithography.

screen is scattered by the reflection on the surface of phosphor particles and by absorption of PVA film on the phosphor particles. Because the scattering is much higher than the absorption by the thin PVA film (<1 μm), the penetration depth of the UV light is primarily determined by scattering. Scattering in the screen extends equally in the vertical (thickness) and horizontal (parallel to substrate) directions. When vertically scattering UV light reaches the substrate underneath the phosphor screen, the PVA film on the phosphor particles and on the substrate undergoes a photoreaction. Consequently, the patterned phosphor screen adheres to the substrate. The minimum size of a patterned phosphor screen adhered to the substrate is calculable for the ideal PVA phosphor screen [120].

Figure 5.10 shows a calculation model for patterned phosphor screens adhered to the faceplate. The pattern size D (strip width or dot diameter) is given by

$$D = D_L[1 + 2\varphi(N_L - 1)/D_L] \tag{5.6}$$

where D_L is the spot diameter of incident UV light, corresponding to the aperture size of the shadow mask. The minimum pattern size of the phosphor screen ($N_L = 1.5$) is

$$D = D_L + \varphi \tag{5.7}$$

If the pattern size of the adhered screen is 30% larger than D_L (e.g., D = 1.3 D_L), the pattern size is acceptable for production. We can calculate the conditions for $2\varphi(N_L - 1)\cdot D_L^{-1} < 0.3$. For color CRTs in TV application, D_L is 200 μm. As $N_L = 5$ and $\varphi = 4$ μm, $2\varphi(N_L - 1)\cdot D_L^{-1} = 0.12$ (< 0.3) and D = 240 μm.

For ordinary color CRTs in TV sets (267 × 263 pixels), the screen layers and particle sizes are not sensitive to the preparation of the patterned phosphor screens. The screen can be patterned on the substrate for TV applications without difficulty, although the screens have defects containing some amount of the clumped particles (less than 30 particles = $3^3\varphi$). Thus, phosphor screens with defects have been produced for 50 years. However, there is difficulty in the production of high-resolution CRTs for VTDs.

The production of a high-resolution screen (1280 × 1024 pixels for XVGA), in which D_L is typically 50 μm, is a different story. For $N_L = 5$ and $\varphi = 4$ μm, we have $2\varphi (N_L - 1) \cdot D_L^{-1} = 0.6$ (>0.3) and D = 80 μm (= 50 × 1.6 μm), which are not accepted in practice. For obtaining a smaller value of $2\varphi (N_L - 1) \cdot D_L^{-1}$ (>0.3), there are alternative choices for either N_L or φ. Using a phosphor powder of $\varphi = 4$ μm and $N_L = 2$, we obtain $2\varphi (N_L - 1) D_L^{-1} = 0.16$, and D = 58 μm, which is acceptable. If $N_L = 5$ and $\varphi = 2$ µm, we obtain $2\varphi (N_L - 1)D_L^{-1}$ = 0.32 and D = 66 μm, which is not acceptable. CRT manufacturers must alter the operating conditions of their screening facilities for evenly screening phosphor powders with two layers of particles.

Because high-resolution screens have a narrow separation between patterned screen pixels, the scattering of emitted CL light in the screen should be minimized in patterned screen pixels. With a thick screen, scattered CL light gets into neighboring phosphor screen pixels in different colors. The scattering of CL light to neighboring phosphor screen pixels smears the images. The minimization of CL light scattering to neighboring screen pixels occurs with two layers of particles. It is therefore preferable to screen the large particles (4 μm) with two layers of particles evenly. Rigid control of the number of layers of phosphor particles, rather than rigid control of the average particle size, is necessary for the production of high-resolution phosphor screens that display clear images on phosphor screens. When the number of layers of phosphor particles is well controlled within $N_L = 2.0 \pm 0.2$ evenly, the patterned phosphor screen with high resolution can smoothly produce particles of 4 ± 2 μm.

When one has a screening technology of phosphor powder with two layers of particles evenly, like the screen shown in Figure 5.3, a best phosphor powder is the polycrystalline particle in a narrow distribution [21]. We do not recommend the spherical particles for screening with a poor adhesion on substrate. A high-resolution phosphor screen (>1000 pixels cm^{-2}) is made from polycrystalline phosphor powder. A high-resolution blue phosphor screen is made from particles that distribute in a narrow sizes ($\varphi = 4$ μm). Figure 5.11 shows a microscope picture (200X) of the ZnS:Ag:Cl blue phosphor. The screen for $N_L = 2$ is prepared with a long settling time and a slow drying rate. The dried screen similar to that shown in Figure 5.3. Striped screens are produced with the screens by application of photolithography to the dried screens in various doses. The patterned screens solidly adhere to the substrate. Consequently, the pattern sizes by the given exposure do not change with the developing conditions. Figure 5.12 shows the sizes of the patterned screens after development. The patterned size is linear with

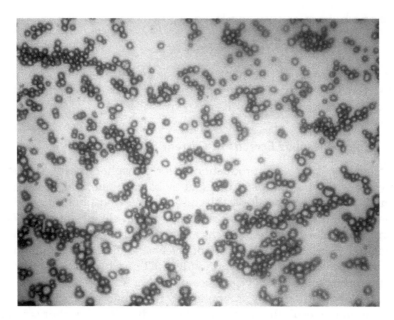

FIGURE 5.11
Photo (300X) of ideal phosphor particles for a defect-free screen.

the exposed doses (arb.) and the Bunsen-Roscoe' law is applicable to the curve. The results definitely show that if one has the ideal phosphor screen, then one can calculate the threshold adhesion conditions of the patterned phosphor screens on the faceplate. The sizes and edges of the patterned screens are not influenced by the developing conditions.

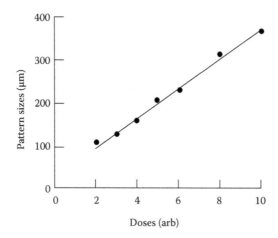

FIGURE 5.12
Pattern sizes of developed PVA defect-free phosphor screens as a function of exposure dose.

5.6   Practical Phosphor Screens

There are two kinds of phosphor screens in practice — monochrome and color screens — which are produced in different ways. We separately discuss these color and monochrome screens.

5.6.1   Color Phosphor Screens in Current Screening Facilities

In present-day color CRT production, color phosphor powders predominantly screen on the faceplate of CRTs by a technique that involves spin coating a PVA phosphor slurry. The use of PVA phosphor slurries is the reason for the dispersion of phosphor particles in slurries of pH $\approx$ 7.0 [5]. Patterning of color phosphor screens on the faceplate is performed by photolithography on dried PVA phosphor screens. Although 200 million color CRTs are produced annually worldwide, CRT producers do not produce ideal phosphor screens. They produce phosphor screens that have defects. Currently, photolithography is applied to these phosphor screens having defects, with the result that it is difficult to study color phosphor screens on the faceplate of CRTs.

We can examine the screening problems of phosphor powders in present-day production facilities using an ideal phosphor powder (Figure 5.11). If the screen is thicker than five layers, a pinhole-free screen is obtained on the faceplate. When photolithography is applied to the screen, there is no screen left on the faceplate after the developing process. This is because the incident UV light does not penetrate through the thick phosphor screen to the faceplate using current recipes for PVA phosphor slurries. If the screen thickness is less than three layers, the screen exhibits numerous pinholes. These pinholes do not emit CL, resulting in a low-luminance screen. Furthermore, large pinholes are generated in the screen by the screening order. When the faceplate already has patterned blue and green phosphor screens, the red phosphor powder finally screens on the faceplate. The red phosphor screen has extra-large pinholes in the center area and a thick screen on the sides, even though the estimated screen thickness is thicker than five layers. The large pinholes are the consequence of neighboring screens sucking water from the wet red screen. Moving water carries unsettled red phosphor particles from the center to the side areas of the screen. Figure 5.13 shows a microscope picture (500X) of the red phosphor screen (by polycrystalline particles) with five layers (estimated). The defects are enhanced by the application of spherical particles. Current CRT producers do not accept the ideal phosphor powder for their production lines because of the numerous screen defects. This makes it difficult for CL phosphor producers to determine what CRT manufacturers will deem acceptable.

Present-day screening facilities in color CRT production use phosphor powders that contain large plate particles and small particles, as shown in

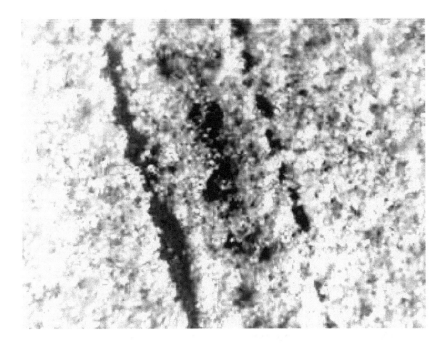

FIGURE 5.13
Photo of large pinholes in a patterned PVA red phosphor screen made from particles of equal size.

Figure 5.14. In addition to phosphor particles, the productivity over time is an important concern in CRT production. To increase the productivity, the wet screens are rapidly dried in high-temperature heaters before the phosphor particles settle on the faceplate. The produced phosphor screen contains defects (i.e., many pinholes and thick spots). Then, screens thicker than seven layers are produced in order to fill in the pinholes. Naturally, Bunsen-Roscoe's law is not applicable to screens having defects. We can analyze the screening mechanisms of the phosphor powder shown in Figure 5.14 [20].

Before analysis of the powder, a problem must be clarified. Commercial phosphor powders sometimes contain clumped particles in large sizes. The clumped particles are due to insufficient removal of the by-products from the produced phosphor powders. The clumped particles cannot be distinguished from primary particles under SEM image observation. The clumped particles are only detected under an optical microscope (<100X) in transmission mode, as the sample is made on the glass plate in thin layers, one to three layers, by a phosphor slurry. The clumped particles are detected as dark spots. When the phosphor powder contains clumped particles, the clumped particles deposit on the center and cross area in the dried screens. Figure 5.15 illustrates the deposited areas of the clumped particles. When the phosphor powder is washed with water heated to 80°C for 10 hours, the clumped particles disappear from the phosphor powder.

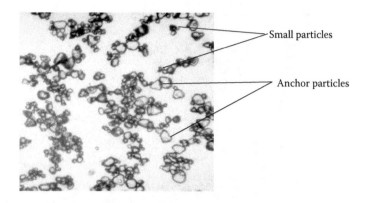

FIGURE 5.14
Photo of particles of commercial blue phosphor powder (300X).

The results indicate that the clumped particles are the responsibility of phosphor producers.

Particular attention should be paid to the role of the large plate particles in Figure 5.14. If the phosphor powder contains these large particles, the large particles quickly settle on the faceplate in the screening slurry, but the particles that are covered with PVA solution do not strongly adhere to the faceplate. If the large particles have flat growing surfaces, the wet flat surface instantly and strongly adheres to the wet faceplate by the aid of a vacuum after some short movement in distance. The strongly adhered particles on the faceplate act as the anchor in the moving slurry. The plate microcrystals, which have flat growing faces, are preferred as the anchor particles. The anchor particles should have a clean surface (without surface treatment) for tight adhesion. Anchor particles should be as large as possible. The size is limited by the adhesion of phosphor particles on the faceplate in the final CRT and

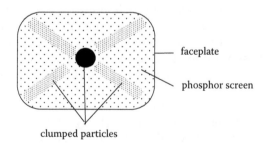

FIGURE 5.15
Schematic illustration of clumped particles deposited on the center and cross-areas in a dried PVA phosphor screen.

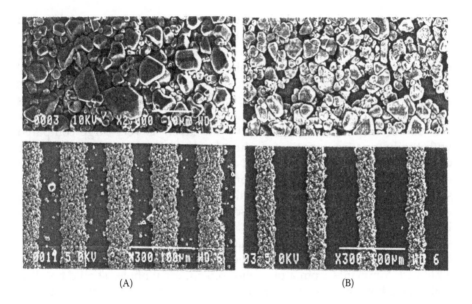

(A)   (B)

FIGURE 5.16
SEM micrographs of particles and patterned phosphor screens: (A) the commercial blue phosphor, and (B) the powder that has removed the large and small particles from (A). (*Source:* Adapted from Reference 3.)

cross-contamination between patterned phosphor screens after the development process. The size is not limited by screening conditions [3].

Figure 5.16(A) shows SEM pictures of particles of commercial blue phosphor powder (above) and patterned phosphor screens (bottom). The patterned phosphor screens are wide. In addition to the width, there are many contaminated particles between the patterned screens. The contaminated particles are very large plate particles. The wetted flat faces of the large particles adhere to the wetted substrate in the development process. The large and small particles have been removed from the commercial phosphor powder by a water sieve technique. Figure 5.16(B) shows the SEM picture of the particles and patterned screen. The screen quality is significantly improved by this removal process. It is clear from Figure 5.16(B) that this is an adequate particle size for anchor particles.

Furthermore, in the CRT production process, the PVA resin covering the phosphor particles bakes out from the screen. Consequently, phosphor particles in manufactured CRTs have no binder. The particles adhere to the faceplate by van der Waals forces, which is weak adhesion. Mechanical shock of the CRT removes the large particles (large mass) from the screen, generating dust in the sealed CRT. The dust in the CRT triggers an arc discharge during CRT operation, and the arc destroys electronic chips in the device. The dust produced in the CRT, and cross-contamination, limit the size of the anchor particles, that is, within double the average particle size (<2 φ) of the phosphor powder, which is much smaller than the plate particles in Figure 5.14.

In present-day screening facilities, the anchor particles control the generation, size, and distribution of pinholes in screens thicker than three layers. Screen luminance will improve by reducing the screen thickness (8% per layer). If the number of screen layers reduces to three layers (average) from seven layers, the screen luminance will increase by 32%. The variation in screen luminance by the layers is very large compared with the luminance variation of phosphor powders (a few percent). The difficulty is that one cannot obtain a generalized phosphor powder; the optimal size and amount of anchor particles differ with the individual screening facilities and drying conditions. The optimal conditions for the anchor particles in thin phosphor screens are empirically determined by individual screening facilities.

Figure 5.14 describes the role of small particles. A phosphor screen thinner than three layers inevitably has pinholes, which do not emit CL. To obtain a high CL luminance from a phosphor screen, the pinholes should be filled with phosphor particles. In a thick screen, the adhesion of the patterned screens onto the faceplate cannot be controlled with anchor particle alone, because thick screens drop out from the faceplate during the development process. We need something that will provide for the adhesion of thick phosphor screens onto the faceplate. This requirement is achieved by the formation of tiny pinholes (10 to 50 μm) [20].

If there are gaps between particles in the screen, the water stream in the wet PVA phosphor screen selectively carries small particles through the gaps to the bottom of the thick phosphor screen. The anchor particles at the bottom of the screen are surrounded by the gathered small particles. The small particles hold a significant amount of water due to capillary action. The small particles hold water up until the final drying step. If the water in the gaps is suddenly vaporized due to rapid heating, the water vapor blows out from the bottom of the phosphor screen, carrying the particles along the path to the surface. Carried particles pile up around the tiny holes. Consequently, the tiny holes are surrounded by a large number of particles (forming a high mountain), like a volcanic mountain with a crater in its center [20]. Hereafter, the tiny pinholes are referred to as crater pinholes. Because there is no water in the dried screen, the crater pinholes are "frozen" on the screen. The crater pinholes are through-holes, as confirmed by microscopic observation in transmission mode.

A further complication involves the generation of crater pinholes — that is, the addition of organic surfactants to the slurry. Historically, the surfactants were added to the slurries to disperse the phosphor particles, with the misinterpretation that the piled-up particles were the clumped particles. In reality, many surfactants in the slurry are involved in gathering the small particles in the wet screen and in holding the water at the small particles.

Figure 5.1 shows these crater pinholes. The white spots are reflected light on the flat, growing face of anchored particles. The valleys (thin screen) between the mountains are formed by a water stream of drying PVA solution. There are many tiny pinholes at the top of mountain, not in the valley, thus proving the mechanism of crater pinholes.

In the production of thicker screens, (1) the size and content of the small particles, (2) the heating conditions for drying the screen, and (3) the kinds and amounts of added surfactants vary with each production facility.

It is clear that in color CRT production, the patterning of color phosphor screens is produced by photolithography on the defective phosphor screens, through the pinholes to the substrate. The incident UV light directly penetrates into the pinholes, and the PVA film on the phosphor particles arranged around the pinholes absorbs the incident UV light for photoreaction to adhere the phosphor particles to the substrate. Figure 5.17 illustrates a model for the patterning of phosphor screens by photolithography on a screen having pinholes.

In Figure 5.17, the black particles represent the photoreacted PVA film on the particles, and the white particles represent the absence of photoreacted PVA film. Because all of the phosphor particles in the patterned screen do not have the photoreacted PVA film, adhesion of the patterned screen onto the faceplate is influenced by the developing conditions, resulting in a poor edge-cut of patterned screens, and partial peeling away of the lined screen and drop-out of the dotted screens. The poor reproducibility has been interpreted of a weakness in photosensitivity. The reality is not the photosensitivity of the PVA film, but rather poor reproducibility in the generation and density of the pinholes. Reproducibility can be improved by excess exposure doses, which extend the pattern size. The extension of horizontal scattering (parallel to the faceplate) of the UV light should be minimized to obtain patterns as small as possible. The pattern size is minimized by high ADC

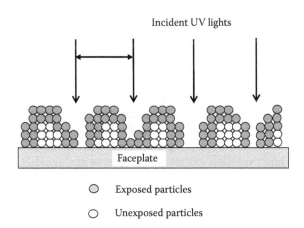

FIGURE 5.17
Schematic illustration of adhesion mechanisms of patterned phosphor screens that have pinholes (p denotes the average distance between pinholes).

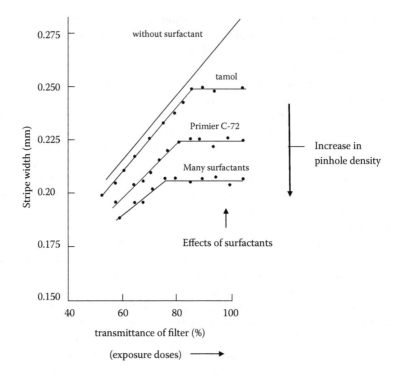

FIGURE 5.18
Effect of the addition of surfactants to a PVA phosphor slurry. Pinholes are generated by the addition of surfactants to the slurry.

concentrations (greater than 2 times that of an ideal phosphor screen). Because the reasons for the poor reproducibility are not well understood, the production of patterned phosphor screens is variously controlled by the (1) generation, (2) size, (3) density, and (4) distribution of pinholes, and the (5) concentration of $HCrO_4^-$.

It was found empirically that the reproducibility of patterned phosphor screen sizes could be controlled by the addition of the multiple surfactants. Figure 5.18 shows the effects of surfactants on PVA phosphor slurries. The density and size markedly change with the addition of surfactants.

Bunsen-Roscoe's law does not apply to an entire screen that is of uneven thickness and has pinholes, but is applicable to the localized area around the pinholes. In this way, the size of the patterned screen adhered on the faceplate (D_{pin}) is given by

$$D_{pin} = D_L + 2p \tag{5.8}$$

where p is the average distance between pinholes. The dose of the UV light is not involved in Equation (5.6), so that D_{pin} is a constant under wide exposure doses. A narrow D_{pin} is obtained with dense pinholes (small p), and high luminance is obtained with small pinholes. For a determination of average p, measurements using an optical microscope are recommended.

The crater pinholes are only generated in thick phosphor screens (more than four layers); they do not form in thin phosphor screens (less than three layers). A suitable color CL phosphor powder for screening should contain anchor particles.

Here is an important message for CRT production facilities where exposed color PVA phosphor screens are developed with city water after removal of small grains (sand), instead of using expensive, deionized (DI) water. Established CRT production facilities use DI water for the development of exposed phosphor screens because ZnS phosphors can become contaminated with the elements of the iron group (i.e., Fe, Ni, Co) present in city water. It was believed that the CL intensity of ZnS phosphors is dramatically reduced (killed) in the presence of such contaminants in the developing process [7]. This is false. Presently, the surface of ZnS:Ag:Cl blue phosphor particles is heavily pigmented with $CoAl_2O_4$ microclusters, and Y_2O_2S:Eu red phosphors are pigmented with Fe_2O_3 microclusters. The pigments directly adhere to ZnS phosphor particles, and the pigments are in direct contact with ZnS phosphor screens in the screening process of phosphor powders. There is no report that the CL intensities of patterned blue and green ZnS phosphor screens have decreased due to contamination by pigments in the screening process. The pigments do not affect the CL intensities or color of the ZnS phosphor screens. Consequently, the exposed phosphor screens on the faceplate can be safely developed using inexpensive city water, after the removal of small grains by filtration.

5.6.2 Monochrome Phosphor Screens

Presently, monochrome CRTs are used in the areas of medical diagnosis in hospitals (x-ray, CT, CDT images), monitoring of professional video camera, security monitors in factories, stores, and homes, light source of printers, image intensifier tube, and image projectors, with a high resolution of images (more than 3 times that of color images). Some monochrome CRTs are still used in low-cost black and white TV sets. The widely used monochrome phosphors (black and white) include (1) white-emitting Y_2O_2S:Tb:Eu phosphors, and (2) a mechanical blend mixture of blue ZnS:Ag:Cl (blue) and (Zn,Cd)S:Ag:Cl (yellow) phosphors. The CL screen in a high-resolution CRT is made from the Y_2O_2S:Tb:Eu phosphor, and black and white TV sets use the blend phosphor powder.

For TV application, the monochrome phosphor screens are traditionally produced on the faceplate of CRTs by sedimentation of phosphor particles from the suspension, which is a mixture of phosphor powder, DI water, and a small amount of K-silicate ($xK_2O \cdot SiO_2$) in the CRT envelope. By adding either $Ba(NO_3)_2$ or $Ba(CH_3COO)_2$ as an electrolyte to the suspension, a large amount of insoluble SiO_2 microclusters (insulators) is instantly produced in the suspension as a result of the chemical reaction between the K silicate and the Ba electrolyte. The produced SiO_2 microclusters are adsorbed onto the surface of suspended phosphor particles, and the phosphor particles in the suspension clump together due to the binding force of -Si-O-Si- chains. Clumped particles in the large sizes (10 to 30 particles) rapidly settle down in the suspension. The SiO_2 microclusters on clumped particles in the screens give a flicker of black and white TV images on the phosphor screen.

After 1990, the information density of monochrome CRTs increased to XVGA (1280×1024 pixels) and medical applications (2560×2048 pixels) from TV applications (267×263 pixels), narrowing the pixel size of images (to 0.16 mm from 1.5 mm in a 20-inch screen). Thus, image quality (flicker, smear, and contrast) becomes an important concern at high luminance (300 to 1500 cd m^{-2}) and at the distance of distinct vision (30 cm away from the screen). The human eye is irritated by flickering images. The removal of flickering images from the black and white screen is an urgent concern with regard to CL screens in high-resolution CRTs. As described in Section 4.4, a black and white flicker-free phosphor screen can be produced without K silicate in the phosphor suspension.

6

The Production of Phosphor Powders

Previous chapters have revealed the (1) optimization of the generation of cathodoluminescence (CL) in phosphor particles, (2) optimization of screen luminance, (3) optical and electrical properties of phosphor particles under irradiation of electrons, and (4) optimization of screening conditions of phosphor powder on the faceplate of CL display devices. Although we have a good understanding of the items described, we may not have sharp and clear CL images on phosphor screens in CL display devices concerning the commercial phosphor powders currently in use. The production processes of commercial CL and PL phosphor powders have been established by empirical skills, but we do not yet understand the details of the production process scientifically. Commercial phosphor powders are far from ideal; we have yet to design a production process for highly optimized phosphor powders for practical use. Chapter 6 discusses the complications encountered in the production of CL phosphors. The discussed results are also applicable to PL phosphor production.

Commercial CL phosphor powders have been produced for more than 50 years, and annual, worldwide production volumes of CL phosphors total about 4000 tons. The details of CL phosphor production remain vague. Despite this vagueness, every one produces brilliant CL phosphors using raw materials of luminescent grade.

The major raw materials employed for host crystals are the same. Although different preparative techniques (e.g., sol-gel, organometallic reaction, hydrothermal synthesis, co-precipitated reaction, etc.) of the raw materials of host crystals have been proposed, the resulting CL properties are the same, indicating that the variation does not derive from the raw materials. The produced phosphor powders using the raw materials generated by various techniques emit similar CL intensity and color. For this reason, the preparation technique does not absorb the attention of the phosphor producers, because these techniques are inadequate for the production of a "screenable" phosphor powder. The phosphor producers focus their attention on screenable phosphor powders, rather than the CL intensities and the color of phosphor powders.

Phosphor producers prepare phosphor powders by heating a blend mixture of powdered raw materials (i.e., solid reactions). The ZnS:Ag:Cl blue

TABLE 6.1

Recipe for Blend Mixture of a-ZnS, NaCl, and S for ZnS:Ag:Cl Blue
Phosphor Production (Traditional)

Material	Weight (g)	Percentage (wt. %)	Molar Ratio
a-ZnS (containing 10^{-4} Ag)	3400	100	1.00
NaCl	53	1.6	3×10^{-2}
S	70	2.1	7×10^{-2}

phosphor, as an example, is produced as follows [7]. The recipe in Table 6.1
mechanically blends in a V-shaped blender. The blend mixture charges in a
quartz crucible (500 ml), and then the crucible is covered with a lightweight
coverplate. The crucible is put directly into a box furnace heated to above
800°C. The phosphor particles grow in the heated blend mixture to the desired
sizes at the temperature of the furnace. Currently, many phosphor producers
use a kiln (continuous) furnace with the same heating concept of crucibles,
but without considering the chemical reactions in the heating mixture. They
focus on protecting the crucible from thermal shock, and use large quartz
crucibles. The particles grown in the crucible are minimally sintered with
solidified materials. The sintered material has been identified as NaCl, under
the assumption that melted NaCl has the flux action for the growth of ZnS
particles. Then one must also consider the removal of NaCl from the heated
products. According to chemistry handbooks, the solubility of ZnS is 68 mg
in 100 ml water, and that of NaCl is 36 g in 100 ml water. Taking the large
difference in solubilities in water into account, one can wash the heated products with deionized water several times to remove NaCl from the phosphor
powders. Completion of the wash is determined by a drop of $AgNO_3$ solution
added to the rinse water (in some cases, the pH value of the rinse water is
used). If there are Cl$^-$ ions in the rinse water, AlCl gels (white) will form in
the rinse water. The washed phosphors dry in a heated oven. The dried
ZnS:Ag:Cl phosphors emit brilliant CL luminance and color, with a variation
of ±15%. The difficulties encountered are not related to CL properties. The
most challenging attribute is the screenability of the produced phosphor powders on the faceplate of CRTs. Each phosphor producer offers its manufactured
phosphor powders to CRT producers for a preliminary evaluation of the
screening test. The trial-and-error approaches used in preliminary screening
tests generate the "in-house secret" of phosphor production.

It was initially thought that screenability related to particle size. The particles
of the produced phosphors widely distribute in the sizes and shapes, such as
the commercial phosphor in Figure 2.13. It was thought that the wide distribution was caused by an uneven distribution of the raw materials in the blend
mixture. The distribution of produced phosphor particles does not change
with the blending time of the mixture, nor with the raw a-ZnS powders. This
chapter provides details regarding the scientific difficulties encountered in
phosphor production. One item often overlooked concerns the by-products
that result as a consequence of chemical reactions in heated products.

6.1 Required Purity of Raw Materials for Phosphor Production

Luminescence intensities are influenced by two kinds of the impurities in CL phosphors: (1) the impurities that influence the transition probability of the luminescence centers, and (2) the impurities that form the recombination centers of EHs, which act as a competitor with luminescence centers in the recombination of generated EHs in particles. The transition probability of the luminescence centers is influenced by electrostatic and magnetic interactions of impurities. The magnetic interactions occur with impurities that within a distance of ten lattice sites of the luminescence center; and the electrostatic interactions occur in the volume within 10^3 lattice sites. The impurities that form the competitive recombination centers are multivalent transition elements.

Because the impurities are less than 1/100 of the concentrations of the luminescence centers (e.g., activators), the variation in luminescence intensity by the impurities is less than 1% — that is, within the measurement error of luminescence intensity. In the purification of raw materials, the impurities are removed from the raw materials to a negligible level. For example, impurities in the raw materials of ZnS phosphors (100 ppm activator concentration) should be less than 1 ppm (<6N). Activator concentrations of Y_2O_2S phosphor are approximately 10^{-2} mole fraction. Y_2O_2S phosphors can accept raw materials of 4N purity. In the early days of phosphor production (i.e., prior to 1970), purification of the raw materials was a major subject of study for obtaining brighter phosphors. Highly purified raw materials (luminescent grade) are currently available on the market.

The remaining impurity in luminescent properties is not an impurity in the raw materials. It is the oxygen that forms additional recombination centers in ZnS phosphors, and that changes the valency of activators (e.g., RE^{3+} in Y_2O_2S) in phosphors.

6.2 Source of Oxygen Contamination

Phosphor producers have difficulty controlling the small variations in luminescence intensities and color of luminescent phosphors. They know that the variation comes from oxygen contamination. It was thought that the source of oxygen contamination was the air in the furnace and/or the air in an open space between charged blend mixture and the crucible. Accordingly, phosphor producers tried to eradicate the oxygen contamination by the following methods: (1) the air in the furnace was substituted by N_2 flowing gas, and (2) active charcoal powder was put on the blend mixture in the crucible. However, unsatisfactory results were obtained.

The blend mixture in powder form inevitably contains many air bubbles (pores). The air in the heating blend mixture in the crucible has been commonly overlooked by phosphor producers. When a crucible is heated, the pores hold a positive pressure of air (and other gases) against an open space between the charged blend mixture and the crucible. With the positive pressure, the diffusion of air from the open space into the charged mixture is difficult. It is also difficult to remove the air from pores in the mixture to the open space of the furnace. The pores in the heating mixture hold the air during the heating process. The positive air pressure in the pores is the reason that both the active charcoal powder on the charged mixture in the crucible and N_2 gas flow in the furnace did not work for the eradication of oxygen contamination in the produced phosphors. The source of oxygen contamination is surely air in the pores of the blend mixture charged in the crucible.

The amount of air bubbles in the charged mixture in the crucible is calculable [26]. The following calculation is made using a blend mixture of amorphous ZnS (a-ZnS). The results are applicable to the blend mixtures of other raw phosphor materials. The density of cub-ZnS crystal is 4.1 g cm^{-3}. The density of the lightly charged mixture in crucible is 0.5 g cm^{-3}. The air occupies 88% volume of crucible [= 1 − (0.5/4.1)], which charges the blend powder. In the densely charged mixture (1.0 g cm^{-3}), air occupies about 75% volume, showing that the volume of air in the charged mixture is less dependent on the charged density of the blend mixture. We can calculate the amount of oxygen in the charged mixture in a practical crucible. If the crucible of 2 liters is fully filled with the lightly charged mixture (10 mole a-ZnS powder with 0.5 g cm^{-3}), the calculated volume of air is 1760 cm^3. Because the air is composed of N_2 (78%) and O_2 (21%), the O_2 gas is 360 cm^3 [= 1760 × (21/78)]. The volume of 1 mole gas is 22.4 × 10^3 cm^3 at room temperature (RT). The charged mixture in the 2-liter crucible contains 0.02 mole O_2 gas {= 360 × (22.4 × 10^3)$^{-1}$}, corresponding to 2 × 10^{-3} mole per mole a-ZnS powder in the charged mixture — higher than the activator concentrations (~1 × 10^{-4} mole per mole a-ZnS).

Oxygen smoothly diffuses into crystallized ZnS:Ag:Cl (and ZnS:Cu:Al) particles; and once diffused, oxygen does not diffuse out from ZnS lattice with the stability of the lattice distortion. The oxygen in pores must be eliminated from the charged mixture before the growth of ZnS particles. ZnS particles grow at temperatures above vaporization of ZnCl$_2$ (T_m = 732°C).

6.3   Eradication of Oxygen from Heating Mixture

At temperatures below 732°C, there are two ways to eliminate oxygen in the pores of the blend mixture in a heating crucible: (1) the formation of metal oxide (solid), and (2) the formation of a harmless SO$_2$ gas. The formation of metal oxide has been empirically found by some phosphor producers, but

TABLE 6.2

Recipe for Blend Mixture of a ZnS Blue Phosphor that Eliminates
Oxygen by Formation of Al_2O_3

Material	Weight (g)	Percentage (wt. %)	Molar Ratio
a-ZnS (containing 10^{-4} Ag)	3400	100	1.0
NaCl	5.3	0.15	2.6×10^{-3}
$AlCl_3 \cdot 6H_2O$	17.2	0.5	2×10^{-3}
S	34.0	1.0	1×10^{-2}

they have not established the elimination mechanisms involved. We first
discuss the details of eradication by the formation of metal oxides.

When the blend mixture contains an appropriate amount of uniformly
distributed $AlCl_3$ fine powder, and when the mixture in the crucible heats
from RT, $AlCl_3$ powder reacts with oxygen at below 400°C to form preferentially Al_2O_3 particles (white) before formation of Al_2S_3 (light yellow). The
sizes of the converted Al_2O_3 (and Al_2S_3) are duplicated with the sizes of $AlCl_3$
particles. Al_2O_3 (and Al_2S_3) particles are chemically stable in sulfur-rich atmosphere, at even 1000°C. Al_2O_3 does not diffuse in the ZnS lattice, and does
not affect the luminescence in ZnS:Cu (or Ag):Cl phosphor particles.

Because 1.5 mole O_2 is consumed to form 1 mole Al_2O_3, the appropriate
amount of $AlCl_3$ is 1.3×10^{-3} mole per mole a-ZnS powder. The empirically
determined optimal concentrations of $AlCl_3 \cdot 6H_2O$ in the blend powder is
2×10^{-3} mole per mole a-ZnS powder [26]. Table 6.2 shows an empirically
optimized recipe for a blend mixture of ZnS:Ag:Cl blue phosphor. Above
5×10^{-3} mole $AlCl_3$, the produced ZnS phosphor powder contains the yellow
particles (Al_2S_3), showing that the excess $AlCl_3$ converts to Al_2S_3. Thus, O_2
gas has effectively been eliminated from the pores in the heating mixture
via the formation of Al_2O_3, if the crucibles slowly heat from RT to 400°C.
Consequently, the CL color of the produced ZnS:Ag:Cl blue phosphor with
the recipe in Table 6.2 falls in y = 0.062 ± 0.002 on x-y color coordinates, and
a decay time of around 10 μs. If the crucible is placed directly into the heated
furnace at above 750°C, oxygen in the mixture can diffuse in grown ZnS
particles, and the complete elimination of oxygen in the mixture will not
occur by the addition of $AlCl_3$ particles.

The ZnS:Cu:Al green phosphor was developed for electroluminescence
(EL) devices in the 1960s. The EL phosphor powder is produced with $3 \times
10^{-3}$ mole Al compounds per mole a-ZnS powder. At present, commercial
ZnS:Cu:Al phosphors for CL applications are produced with 3×10^{-4} mole
Al compounds per mole a-ZnS powder [15]; as a consideration of co-activator, that is not large enough for the eradication of O_2 gas (2×10^{-3} mole per
mole a-ZnS powder) in heating mixture. Furthermore, the crucibles are
placed directly into the heated furnace at above 750°C. Consequently, the
commercial green phosphors are somewhat contaminated by oxygen, giving
rise to 100-μs decay times and a variation in CL intensity and color by lots.

When eradication of the oxygen is made by Al_2O_3, a problem arises in the usage of the phosphor powder. The formed Al_2O_3 microclusters (2×10^{-3} mole per mole ZnS) adhere to the surface of the grown ZnS particles. Al_2O_3 is a chemically stable material, and the Al_2O_3 microclusters cannot be removed from the surface of grown ZnS particles by chemical reaction. Al_2O_3 microclusters remain on the surface of produced phosphor particles. The Al_2O_3 microclusters (insulator) on phosphor particles generate flicker images on phosphor screens in CRTs. The amount of Al_2O_3 (2×10^{-3} mole) per mole ZnS is large enough to generate the flicker. Al_2O_3 microclusters should be removed from the surfaces of ZnS phosphor particles.

There is a sophisticated way to remove Al_2O_3 microclusters: segregation of the melted Na_2S_4 in the heated mixture. When the heated blend mixture contains a small amount of Na_2S_4, the melted Na_2S_4 segregates in the heated products in the crucible due to a temperature gradient. The segregating Na_2S_4 carries fine Al_2O_3 particles. When the amount of segregated Na_2S_4 becomes large enough for the conversion of Al_2O_3 to Al_2S_3, the Al_2O_3 in segregated Na_2S_4 (in a high concentration) converts to Al_2S_3. Al_2S_3 is soluble in an acidic solution. Very delicate conditions, which must be controlled during production, are involved in the segregation of Na_2S_4 and conversion of Al_2O_3 to Al_2S_3 [26]. These delicate conditions include:

1. The soft charge (0.5 g cm^{-3}) of the blend mixture in the heating crucible. The softly charged mixture gives the appropriate pore sizes for segregation of Na_2S_4. Because the pores also work as heat resistance, the heated mixture has a temperature gradient in the cooling process.

2. The Na_2S_4 concentration is approximately 3×10^{-3} mole Na_2S_4 per mole a-ZnS. Above 5×10^{-3} mole per mole a-ZnS, the heated product in the crucible is minimally sintered by the solidified Na_2S_4, indicating no segregation of melted Na_2S_4. Below 1×10^{-3} mole, the amount of melted Na_2S_4 is insufficient to carry the Al_2O_3 particles.

3. The Na_2S_4 concentration at the segregated area must be higher than the threshold concentration for the conversion by alkaline fusion.

4. Al_2O_3 particles should be as small as possible for carrying by segregating Na_2S_4 in a thin layer.

5. Slow cooling rate of the heated crucible ($1.3°C$ per min.) for the segregation of melted Na_2S_4 in a high viscosity mixture.

By controlling these five items, the segregated area in the heated mixture should contain the converted Al_2S_3 (light yellow) and the concentrated and solidified Na_2S_4 (yellow). When the produced phosphor powder is suspended in stirring DI water, and then the suspension is poured onto the sieve, the large yellowish particles deposit on the 140 mesh sieve. An amazing result is obtained by examining these yellowish particles [26].

Figure 6.1 shows an SEM micrograph of the yellowish particles. They are microcrystals with flat growing faces and exhibit a slow crystal growth rate.

FIGURE 6.1
SEM micrograph of residual yellow particles (320X) of ZnS:Ag:Cl phosphor powder.

Some crystals are heavily covered with sintered materials that contain small (4 µm) particles. The surface of a hexagonal crystal on the right-hand side in Figure 6.1 was examined by the electron probe micro-analyzer (EPMA).

Figure 6.2 shows the results: (A) shows the SEM micrograph, and (B), (C), and (D) are the images of Zn, O, and S elements, respectively. Examination by x-ray diffraction shows the strong, weak, and very weak lines, which respectively identify hex-ZnO, cub-ZnS, and Na_2S_4, with a trace amount of $Al_2(SO_4)_3$. The large crystal in Figure 6.1 is, amazingly, hex-ZnO, contrary to the anticipation of the sintered Al_2S_3. There is no report of the growth of ZnO particles (>50 µm) from the heated product of ZnS phosphors in crucibles. To explain the growth of ZnO microcrystals from a sulfur-rich atmosphere, a proposed model is presented below [26].

The conversion from Al_2O_3 to Al_2S_3 in melted Na_2S_4 releases active oxygen. The active oxygen immediately reacts with ZnS to form a ZnO nucleus. When the active oxygen continuously generates at the same place by the segregating Na_2S_4, which caries the fine Al_2O_3, the ZnO crystal is continuously grown to the size of the microcrystals (~100 µm) in the heated mixture. Pure ZnO crystals are colorless. The obtained ZnO microcrystals are light yellow in color. Many small particles of Al_2S_3 (light yellow and $T_m = 1100°C$) adhere to the surface of a growing ZnO crystal. The grown ZnO crystal includes the tiny Al_2S_3 particles in it, giving the light yellow body color.

The elimination of oxygen by the Al compound requires some complicated conditions that should be controlled in ZnS phosphor production. A much simpler way to eliminate the oxygen has been reported: the use of melted and vaporized S in the heating mixture [121]. The S vapor is reactive with O_2.

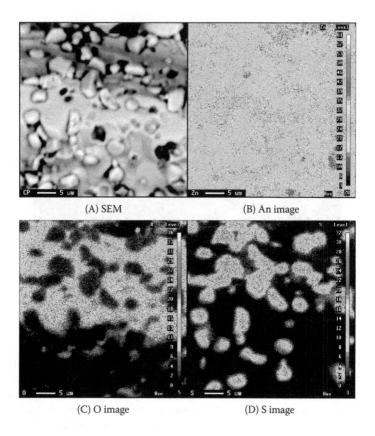

(A) SEM   (B) An image

(C) O image   (D) S image

FIGURE 6.2
EPMA results of residual particles: (A) SEM micrograph, (B) image by Zn element, (C) image by O element, and (D) image by S element.

As the crucible containing the blend mixture (0.5 g cm^{-3}) of Table 6.1 heats from RT to 400°C at a rate of 13°C min^{-1} and maintains it for 90 min, the S (T_m = 144°C and T_b = 444°C) in large sizes (~100 μm) melt in the heating mixture. The charged mixture significantly shrinks the volume by diffusion of the melted S in the pores, similar to a paste made by the addition of water to powder. The density of the charged mixture becomes 1.1 g cm^{-3}. As a consequence, the melted S spreads into the pores and gaps of the blend mixture, and uniformly distributes throughout the heating mixture. When the crucible temperature increases to 600°C from 400°C ($<T_b$ = 732°C of ZnCl$_2$) with a rate of 3.3°C min^{-1}, excess S that has not chemically reacted evaporates in the pores. The evaporated S reacts with O$_2$ in the pores to form SO$_2$, which is harmless to the production of ZnS phosphor particles. Hence, the oxygen in the blend mixture converts to production of SO$_2$ gas in the heating mixture before the start of growth of ZnS particles. Then, the crucible heats up to the temperature at which ZnS particles are grown by the flux action. The same

heat program of the crucibles eradicates the oxygen from the blend mixture of raw materials of Y_2O_2S phosphor powders.

The CL intensities and color of produced ZnS and Y_2O_2S phosphors have a high reproducibility when the blend mixture in the crucible is heated using the programs described above. Remaining problems in phosphor production include the growth of phosphor particles in desired sizes and shapes, which are controlled by (1) flux action and (2) the uniformity of the blend mixture in heating crucibles.

6.4 Identification of Flux Material for Growth of ZnS Particles

The production of ZnS phosphor powder has a long history (more than 50 years). For a long time it was believed that ZnS particles were grown from NaCl in the liquid phase [7, 15, 122]. A low T_m of the alkaline-metal, alkaline-earth, and other metal chlorides has been considered as the substituting flux materials [15, 122]. This conclusion was inferred from experiments employing a crucible with a lightweight coverplate. The flux action of melted NaCl [7] has not been proved scientifically. It is time to revise the flux materials for the growth of ZnS particles based on new findings. Recently, experiments in a sealed quartz capsule, instead of the loosely sealed crucible, have identified the flux material and clarified the growth mechanism of ZnS particles [25].

Commercial a-ZnS powder contains many gases (water, S, air, CO_2, CH_4, and others), which gives the uncertainty of the experiments. A pure a-ZnS powder (nonluminescent) is made by degassing the powder in a high vacuum (10^{-6} torr) at 800°C for 2 hours, in order to avoid the involvement of adsorbed gases in the experiment. The pure a-ZnS powder does not contain any excess S. a-ZnS alone does not grow in the capsule at 950°C. When the capsule contains pure a-ZnS powder and an alkali-metal (Li, Na, and K) chloride, ZnS particles do not grow in the capsule without S. Because the capsule contains the a-ZnS powder and a small amount of $ZnCl_2$ (2×10^{-4} mole per mole a-ZnS), ZnS particles are grown by heat at temperatures above $T_b = 732°C$ of $ZnCl_2$. From the results described, $ZnCl_2$ in vapor is identified as the flux material for the growth of ZnS particles.

ZnS particles are also grown in a capsule that contains pure a-ZnS powder with a small amount of chloride and S. When we take NaCl as the chloride, the chemical reactions in the heated capsule are

$$2NaCl\ (s) + 4S\ (melt) \rightarrow Na_2S_4\ (s) + Cl_2\ (\uparrow) \tag{6.1}$$

$$Cl_2\ (\uparrow) + a\text{-}ZnS\ (s) \rightarrow ZnCl_2\ (s) + S\ (\uparrow) \tag{6.2}$$

where (s), (melt), and ($\uparrow$) represent solid, melt, and gas, respectively. The by-products in the heated mixture are $ZnCl_2$, Na_2S_4, and S, for which T_m is 283,

TABLE 6.3

Physical Properties of the By-Products

By-Products	Body Color	T_m(°C)	T_b(°C)	S_c (g)	S_h (g)
$ZnCl_2$	White	283	732	432	615
Na_2S	White	1180		15	57
Na_2S_4	Yellow	275		s	
Al_2S_3	Light yellow	1100	1150	d	
Al_2O_3	White	2050	2980	i	i

Note: T_m : melting temperature, T_b : boiling temperature, S_c : solubility in 100 ml cold water, S_h : solubility in 100 ml hot water. s, d, and i represent "dissolve," "decompose," and "insoluble" in water, respectively.

275, and 112°C, respectively [123]. Table 6.3 gives the physical properties of the by-products. The chemical reactions of Equations (6.1) and (6.2) are triggered by melted S (>112°C), and $ZnCl_2$ is certainly produced in the heated capsule by chemical reaction. The NaCl in Equation (6.2) is not an essential material for the generation of $ZnCl_2$. One can substitute NaCl by other chlorides of I-a, II-a, III-b, and V-b elements in the Periodic Table. Here we have assigned Na_2S_4, which should be considered a new by-product in the production of ZnS (and Y_2O_2S) phosphors.

The growing mechanisms of ZnS particles in a closed space involve chlorine cycles through $ZnCl_2$ vapor, like the iodine cycles in an iodine lamp [124, 125]. Figure 6.3 illustrates the model of chlorine cycles [4]. $ZnCl_2$ vapor

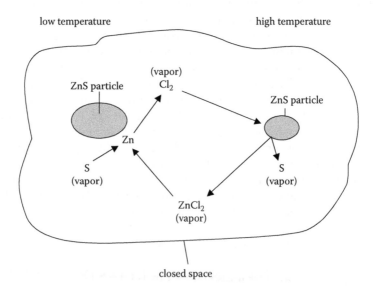

FIGURE 6.3
Model of the growth of ZnS particles by $ZnCl_2$ vapor in a closed space.

TABLE 6.4

Melting Temperature T_m and Boiling
Temperature T_b of Zinc Halides

Halide	T_m (°C)	T_b (°C)
ZnF_2	872	1500
$ZnCl_2$	283	732
$ZnBr_2$	394	650
ZnI_2	446	624 (decomposes)

carries the Zn element to a ZnS particle at the low temperature of a capsule, and then deposits Zn on the surface of the ZnS particle by releasing Cl_2. The released Cl_2 reacts with ZnS at high temperatures to form $ZnCl_2$ by releasing S. The released S reacts with Zn on the surface of the ZnS particle. ZnS particles are grown by chlorine cycles under a temperature gradient.

Table 6.4 gives the T_m and T_b of Zn halides. ZnF_2 has $T_b = 1500°C$, which is too high for ZnS phosphor production (<1000°C). If the blend mixture of ZnF_2 and a-ZnS powder heats at 950°C, the ZnS particles do not grow in the mixture. $ZnCl_2$ has flux action above 732°C. $ZnBr_2$ ($T_b = 650°C$) also has flux action. However, $ZnBr_2$ is not commonly used in ZnS phosphor production because Br⁻ smoothly incorporates into the ZnS lattice [Br⁻ (1.96 Å) and S²⁻ (1.84 Å)]. The incorporated Br⁻ forms an extra luminescence center (Cu-Br) in the ZnS crystal. I⁻ does not diffuse in the ZnS lattice due to the large difference in ionic radii between S²⁻ (1.84 Å) and I⁻ (2.20 Å) [123]. Vaporized ZnI_2 (below 624°C) has flux action.

The preparation of ZnS:Cu:Al green phosphor powder avoids the Cl⁻ incorporation that induces O²⁻ contamination. Hence, the ZnS:Cu:Al phosphor is produced by only vaporized ZnI_2 flux. Here a difficulty arises. ZnI_2 decomposes at T greater than 624°C, and the decomposed Zn and I_2 do not exhibit flux action. By considering the heat conductance of the charged mixture in the crucible, the ZnS particles are grown by raising the furnace temperature to 700°C, and do not grow further at furnace temperatures above 700°C.

6.5 Growth of ZnS Particles in a Crucible

The raw material in ZnS particles is a-ZnS powder. There are two production pathways for a-ZnS powders: (1) precipitation from a $ZnSO_4$ solution by H_2S gas and (2) precipitation by the addition of $(NH_4)_2S$ solution. Many phosphor producers use a-ZnS powder precipitated by H_2S gas. The a-ZnS particles precipitated by $(NH_4)_2S$ are fine particles that do not settle in suspension. Slurries of fine a-ZnS precipitation are dried by application of an atomizer. The dried a-ZnS particles are clumped in sizes of 5 to 10 μm. The size of the clumped a-ZnS particles is controlled by the operating conditions of

(A) clump of a-ZnS   (B) seed ZnS particles

FIGURE 6.4
SEM image of clumped a-ZnS particles and seed ZnS particles (10,000X).

the atomizer. The use of the clumped a-ZnS powders has the advantage of exhibiting a dense charge of the blend mixture in a crucible (~0.8 g cm^{-3}). The dense charge of the blend mixture in a crucible is favorable for the growth of ZnS particles in a narrow distribution. The dense charge of a-ZnS powder precipitated by H$_2$S gas also occurs in a crucible by melted S at 400°C, as already described.

ZnS phosphor particles are grown in the blend mixture in a heated crucible. If the blend mixture contains a sodium compound, and if the crucible is tightly sealed by the coverplate, the mixture in the crucible heated at above 275°C contains appreciable amounts of melted Na$_2$S$_4$ as the by-product. Melted Na$_2$S$_4$ is capable of chemically reacting with the quartz (SiO$_2$) crucible. Therefore, the blend mixture heats in an alumina crucible.

In the process of heating the crucible, a-ZnS particles convert to seed ZnS particles at approximately 650°C. The size of the seed particles (~0.1 µm) is the same for the different precipitation methods of a-ZnS. Figure 6.4(A) shows SEM images of clumped a-ZnS particles precipitated by (NH$_4$)$_2$S, and Figure 6.4(B) shows ZnS seed particles converted from the clumped a-ZnS particles (10,000X). Similar seed ZnS particles (Figure 6.4B) are also obtained with a-ZnS precipitated by H$_2$S gas. The seed particles grow to a desired size (~4 µm) by controlling the flux action in the blend mixture during heating.

It is known that produced phosphor particles do not distribute with a normal distribution, but rather with a log-normal distribution with significance (0.05 %) [5]. Figure 6.5 shows the distributions of seed and grown ZnS particles, as plotted on log-normal graph paper. The seed and grown particles have the same standard deviation as the distribution of the log-normal distribution.

The seal of the crucibles that hold Zn-halide vapor is an important factor for the growth of ZnS phosphor particles. The sealing condition of the crucible can be studied with ZnI$_2$, which has a strong flux action to ZnS.

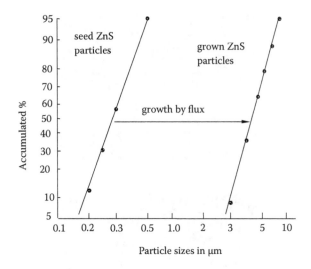

FIGURE 6.5
Size distribution of seed and grown ZnS particles on log-normal graph paper.

6.5.1 Sealing Conditions of Crucibles

The blend mixture of ZnS:Cu:Al with ZnI_2 (Table 6.5) charges in two quartz crucibles in 3 L (liters): (A) loosely sealed crucible with a lightweight cover, and (B) tightly sealed crucible with a heavy cover, and for which contact areas are polished [25]. The crucible (A) is put directly in the heated furnace at 950°C as the reference. This is the established heating of the crucible. The crucible (B) heats from RT to 700°C with a heating rate of 3°C min^{-1} to hold the gases in the crucible. Figure 6.6 shows SEM images of the grown ZnS particles. The sizes of the grown particles markedly differ with crucible type and heating program. The particles grown in crucible (A) are polycrystalline particles of various sizes; the average size is around 4 µm. Particles grown in crucible (B) are extremely large (more than 10 times) compared with particles grown in crucible (A), even with the same blend mixture. The results indicate that holding the ZnI_2 vapor in the crucible is an important concern for the growth of ZnS:Cu:Al phosphors.

TABLE 6.5

Recipe for Improved ZnS:Cu:Al Green Phosphor

Material	Weight (g)	Percentage (wt. %)	Molar Ratio
a-ZnS (containing 2×10^{-4} Cu)	500	100	1.0
$Al_2(SO_4)_3$	0.027	0.005	3×10^{-4}
ZnI_2	1.76	0.44	1.4×10^{-4}
S	2.0	0.2	6×10^{-3}

(A)

(B)

FIGURE 6.6
SEM images of ZnS:Cu:Al green phosphor particles grown by ZnI_2 vapor. (A) Particles grown by rapid heating to 970°C in a loosely sealed crucible, and (B) particles grown by slow heating to 700°C from room temperature in a tightly sealed crucible.

The size and shape of ZnS particles grown in crucible (A) change with the charged density of the blend mixture in the crucible, suggesting that the growth of ZnS particles is not solely controlled by the seal of the crucible. The ZnS particles are grown by flux action in the pores (air bubbles) in the blend mixture. Thus, the sealing of the pores is an important concern in the production of ZnS phosphor powders.

6.5.2 Sealing of Pores by Melted Materials

The mixture charged in the crucible inevitably contains pores. If the pores in the charged mixture are sealed by the melted materials (sealant), the pores in that mixture can hold ZnI_2 vapor. The particle size of ZnS:Cu:Al phosphors surely increases with the addition of the melted materials, such as BiI_3 (T_m = 408°C), and NaOH, which forms Na_2S_4, as a result of chemical reaction with melted S. With a given ZnI_2 concentration, particle sizes of ZnS:Cu:Al particles increase via the use of a sealant to the pores. However, the produced ZnS particles are sintered by the solidified sealant at RT. Because of the

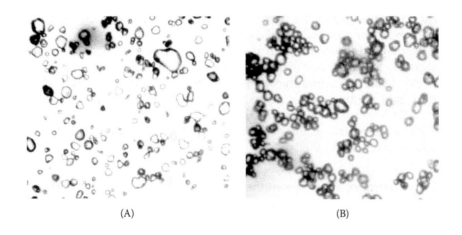

(A)   (B)

FIGURE 6.7
Optical microscope images (500X) of ZnS:Ag:Cl blue phosphor particles produced by (A) softly charged blend mixture (0.4 g cm^{-3}) and (B) densely charged blend mixture (1.0 g cm^{-3}).

complexities involved in using ZnI$_2$ flux with decomposition at above 624°C, the experiments with sealant are performed using ZnCl$_2$.

Equation (6.1) reveals the by-product, Na$_2$S$_4$, which has T_m = 275°C and does not evaporate at even 1300°C. If the pores in the heating mixture are sealed by a layer of melted Na$_2$S$_4$ throughout the heating process, then the ZnS particles grow in the sealed pores. If the crucibles are tightly sealed by a heavy cover, the appropriate amount of NaCl is reduced to 4×10^{-4} mole fraction per mole a-ZnS, as opposed to that of Leverentz's recipe (i.e., 3×10^{-2} mole fraction) (see Table 6.1). At a given heating condition for the crucible, the size of grown ZnS particles corresponds to the amount of ZnCl$_2$ vapor in the sealed pores. The large pores contain a large amount of ZnCl$_2$ vapor, and the small pores contain a lesser amount of ZnCl$_2$ vapor. The softly charged mixture (0.5 g cm^{-3}) contains a wide distribution of pore sizes, and a narrow size distribution is obtained from a densely charged mixture (1.0 g cm^{-3}). Figure 6.7 shows microscope images (600X) of ZnS:Ag:Cl particles, wherein the particles were grown in (A) softly (0.5 g cm^{-3}) and (B) densely (1.0 g cm^{-3}) charged blend mixtures in the crucibles, respectively.

The seal of the pores by melted Na$_2$S$_4$ is fragile, especially at high temperatures. The fragile seal is strengthened by a large amount of Na$_2$S$_4$ (a few mol% per mole a-ZnS), which generates the problems associated with sintering of the grown ZnS particles by solidified Na$_2$S$_4$ at RT. When the crucible is tightly sealed by the coverplate, the produced ZnS powder has a light brown body color due to Na$_2$S$_4$ (yellow). If the crucible is loosely sealed, the S vapor escapes from the heated mixture and Na$_2$S$_4$ converts to colorless Na$_2$S (Table 6.3), for which T_m = 1180°C. It has been found empirically that the large and sealed pores are made by rapidly heating the 500-ml crucible, and the crucible should be loosely sealed by a lightweight coverplate in order to bleach the brown body color. Only quartz crucibles respond to these heating requirements. This is the production process established by Leverentz [7].

If the by-products do not melt at 950°C, they form microclusters (~0.2 μm) that adhere to the surface of grown particles. The microclusters on phosphor particles lessen the sintering of the produced ZnS powder. Typical raw materials of the by-product microclusters are II-a halides, such like as Ca, Sr, and Ba halides. The by-products have T_m greater than 1000°C. The complexity of the control of the pore sizes, body color, and sintering of heated products gives rise to the so-called "in-house secret."

6.5.3 Control of Pore Size in the Heating Mixture

ZnS phosphor particles should be grown in the sealed pores in equal size. The pores in softly charged blend mixtures distribute over a wide range of sizes. This results from the dense charge of the blend mixture in the crucible using the clumped a-ZnS powder precipitated by $(NH_4)_2S$. Equalization of the pore sizes in the charged mixture is achieved in a simple way, even with a-ZnS powder precipitated by H_2S gas [25]. The blend mixture of the recipe in Table 6.6 softly charges (0.5 g cm^{-3}) in an alumina crucible. When the crucible heats from RT to 400°C ($<T_b(S) = 444°C$) with a heating rate of 13°C min^{-1} and maintains the crucible at 400°C for 90 min, S particles (~100 μm) melt. The melted sulfur spreads into the pores and gaps between a-ZnS particles, resulting in a uniform distribution of S throughout the heating blend mixture. The spreading of melted S gives rise to a decrease in the volume of the charged mixture (from 0.5 g cm^{-3} to 1.0 g cm^{-3}). An appropriate amount of S is required for this volume decrease — greater than 1 wt. % a-ZnS powder. As a result of the volume decrease of the blend mixture, the pores distribute in a narrow range of the sizes. The sizes of the grown particles at 970°C certainly distribute in a narrow range, as shown in Figure 6.7(B). The average size is 2 μm, which is half the desired size (4 μm). The maximum heating temperature of the furnace is 970°C; that is, it is limited by the transition from cubic to hexagonal ZnS particles. At the given heating T, there are two ways to grow large particles: (1) increase the Zn halide vapor in the pores using a tightly sealed crucible, and (2) increase the pore size.

For a given crucible, the pore size can be increased by thermal expansion of the gases in the pores. The expansion of the gas in the pores is achieved by the addition of water to the blend mixture [121]. Thus, 100 g a-ZnS powder (taken from the 3400-g powder in Table 6.7) is placed in a 500-ml beaker.

TABLE 6.6

Recipe for Improved ZnS:Ag:Cl Blue Phosphor

Material	Weight (g)	Percentage (wt. %)	Molar Ratio
a-ZnS (containing 10^{-4} Ag)	3400	100	1.0
NaCl	1.5	0.06	4×10^{-4}
S	70	2.1	7×10^{-2}

TABLE 6.7

Recipe for ZnS:Ag:Cl Phosphor Particles (4 μm, average) by the
Addition of Water

Material	Weight (g)	Percentage (wt. %)	Molar Ratio
a-ZnS (containing 10^{-4} Ag)	3400	100	1.0
NaCl	1.5	0.06	4×10^{-4}
S	70	2.1	7×10^{-2}
water	34	1.0	5×10^{-2}

After the addition of 34 g DI water (1 wt. %) to the powder, the wet powder
is well mixed using a glass bar. The wet powder then transfers to the original
powder, and the powder is blended by a V-shaped mill. The blend mixture
softly charges in the alumina crucible. The crucible heats in the furnace
operating using the previous temperature program. The average size of the
produced particles is 4 μm, revealing the expansion in pore size. The distribution and size of the pores in the heating mixture can be controlled by the
amount of added water.

6.5.4 Growth of Plate Particles

The phosphor powder in Figure 6.7(A) contains many large plate particles.
For screening phosphor powder in present-day screening facilities, some
amount of plate (anchor) particles is essential. The following experiments
might explain the growth of plate microcrystals in the pores. The small seed
particles (~1 μm) blend in the mixture of raw materials. The blend mixture
densely charges in the crucible. The produced phosphor powder contains
some amount of plate particles (microcrystals), and the number of plate
particles in the produced powder changes with the amount of added seed
particles. At the given heating conditions, the size of the plate particles
changes with the ratio of the seed particles to a-ZnS powder (seed:a-ZnS);
the smaller the ratio, the larger the plate particles.

In the experiments in Figure 6.7, a large pore may have a seed particle
during early heating using a large temperature gradient. If the flux in the
pore preferentially acts on the seed particle, the seed particle in the large
pore grows along a selected growing axis that gives rise to a plate particle.
If the densely charged mixture does not contain seed particles, there is no
plate particle in the produced phosphor powder, even though the particles
are grown to large sizes (10 times), as shown in Figure 6.7(B).

6.5.5 The Size and Shape of Crucibles

The amount of gases resulting from chemical reactions in the heating mixture
should be minimized. The minimum amount of S for elimination of O_2 in
the heating mixture is 2×10^{-3} molar fraction per a-ZnS. The Cl consumed

TABLE 6.8

Recipe for Improved ZnS:Ag:Cl Blue Phosphor

Material	Weight (g)	Percentage (wt. %)	Molar Ratio
a-ZnS (containing 10^{-4} Ag)	2000	100	1.0
$MgCl_3 \cdot 6H_2O$	2.5	0.1	3×10^{-4}
S	64	3	1×10^{-1}
Water	20	1.0	5×10^{-2}

as co-activator is 1×10^{-4} mole per mole a-ZnS powder and the chlorine cycles for growth of ZnS needs 2×10^{-4} mole chloride per mole a-ZnS. The minimum amount of Cl in the heating mixture is 3×10^{-4} mole per mole a-ZnS. Table 6.8 gives the recipe for a mixture in a tightly sealed crucible. To avoid sintering by the melted Na_2S_4, the recipe in Table 6.8 does not contain an Na compound, and the raw material for $ZnCl_2$ is $MgCl_3 \cdot 6H_2O$.

The volume of the generated gases at 1000°C is estimated at 2×10^3 cm³, which is one-tenth of 1 mole gas under 1 atm and RT. Although the pressure of the generated gases in the crucible is less than 1 atm, the crucible configuration is an important concern for minimization of the pushing pressure to the cover by the Zn-halide vapor. The pushing pressure (P_{cover}) of the Zn-halide vapor to the cover is given by

$$P_{cover} = RT \, V^{-1} \tag{6.3}$$

where R is the gas constant, V is the volume of the crucible, and T is the temperature in Kelvin (K). At a given T and with a given heating mixture, the pushing pressure can be reduced by the use of a large crucible (V). At a given V, the pushing pressure to the cover per unit area is reduced using a wide crucible. Therefore, one should use a crucible that is as large and as wide as possible (larger than 3 liters), and a heavy coverplate is preferred for tightly sealed crucibles in ZnS phosphor production. This may provide an explanation for why a laboratory sample (using crucible of less than 50 ml) always has poorer quality compared with a commercially produced sample.

6.6 Heating Program of Furnace for ZnS Phosphor Production

The crucible, in which the blend mixture charges, heats in the furnace to grow the phosphor particles to the desired size. A reliable phosphor powder is produced by heating the blend mixture from RT according to the proper heat program. A good heat program for the recipe in Table 6.8 is given below: The blend mixture softly charges in an alumina crucible. A thin alumina plate, the size of which is slightly smaller than the inner diameter of the crucible, is

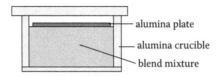

FIGURE 6.8
Suitable crucible that charges the blend mixture for ZnS phosphor production.

placed on the charged mixture. The alumina plate partitions the charged mixture from the large space between top of the mixture and the cover, for which the $ZnCl_2$ vapor has a different flux action. Figure 6.8 illustrates a crucible in which an alumina plate is placed on the charged mixture.

The crucible heats from RT to 400°C at a rate of 13°C min^{-1} and maintains that temperature for 90 min. Then the crucible heats to 600°C at a rate of 3.3°C min^{-1} and maintains that temperature for 30 min. As the temperature rises, some amount of S does not chemically react with the added compounds in the blend mixture. That excess S evaporates from the melted S. The slowly evaporating S reacts with the O_2 in pores to form SO_2. The excess pressure from the gases generated in the crucible pushes up the cover and the gases escape from the crucible. The crucible then heats up to 950°C at a rate of 3°C min^{-1}. Above 732°C, the $ZnCl_2$ in the pores evaporates, and the $ZnCl_2$ vapor serves as the flux in the growth of ZnS particles in the pores. In the case of $ZnCl_2$ flux, the growth of ZnS particles is suppressed by slow heating rates (e.g., 1°C min^{-1}) with the small temperature gradient in the pores. Figure 6.9 shows the grown particle sizes (average) with heating time at 950°C. The temperature of the total heating mixture in the crucible reaches

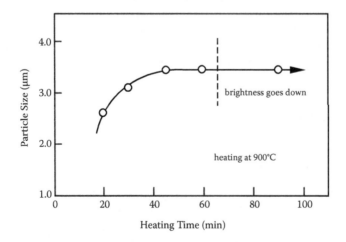

FIGURE 6.9
Particle sizes (average) of ZnS:Ag:Cl particles as a function of heating times at 950°C.

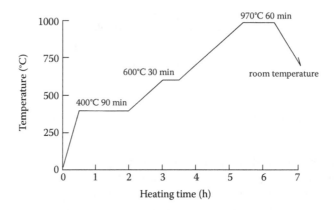

FIGURE 6.10
Heating program of furnace for the production of ZnS:Ag:Cl blue phosphor powder in a tightly sealed crucible.

equilibrium with the furnace temperature for 45 min heating (average), and average ZnS particles no longer grow with additional heating time. The time needed to reach equilibrium varies with crucible size and the arrangement of heaters in the furnace. The ZnS particles that distribute in a narrow size range are produced for a time longer than that required for equilibrium. The crucible remains at 950°C for 60 min. Then, the power to the furnace is turned off so that the crucible can cool to RT in the furnace. Figure 6.10 shows the improved heat program of ZnS:Ag:Cl phosphor powder using $ZnCl_2$ flux.

With the ZnI_2 flux, the heat program somewhat differs from that in Figure 6.10. Figure 6.11 shows the heat program for the recipe of ZnS:Cu:Al green phosphor powder with ZnI_2 flux (see Table 6.5). The particles of ZnS:Cu:Al phosphor grow during the period of increasing temperature between 500 and 700°C. The grown particles have a brown body color due to I_2, which escapes from the crucible at 950°C. The heating conditions may somewhat change with the particular crucible and furnace used.

6.7   Removal of By-Products from Produced ZnS Phosphor Powders

It has been shown that the traditional production of ZnS phosphor utilizes sealed pores from melted Na_2S_4 to grow ZnS particles according to the recipe given in Table 6.1 [7]. At 950°C, the melted Na_2S_4 preferentially penetrates into the small gaps between the grown particles via capillarity, especially the contact gaps between the flat growing faces of ZnS particles and the gaps between small particles. When the crucible cools down from 950°C, the $ZnCl_2$

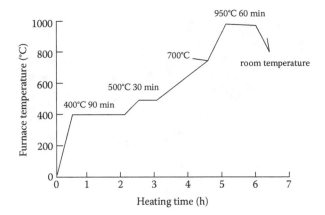

FIGURE 6.11
Heating program of furnace for the production of ZnS:Cu:Al green phosphor powder in a tightly sealed crucible.

vapor liquefies at temperatures below T_b and liquefied $ZnCl_2$ diffuses in the melted Na_2S_4. The amount of $ZnCl_2$ (3×10^{-4} mole maximum) is negligibly small compared with melted Na_2S_4 (a few percent). The ZnS particles in the crucible are sintered by solidified Na_2S_4, including small amounts of other by-products. The solidified Na_2S_4 in the narrow gap between flat growing faces tightly binds the particles. Figure 6.12 shows an SEM image of bound ZnS particles as grown by solidified Na_2S_4. Surfaces of the particles are covered with a film of solidified Na_2S_4. Small particles on the surface of the particles are the particles of other by-products. The by-products on the surface of grown particles should be completely removed from the phosphor powder prior to use.

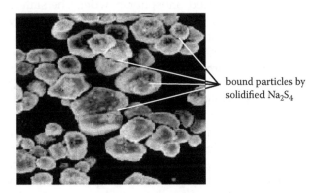

bound particles by solidified Na_2S_4

FIGURE 6.12
SEM image (2000X) of bound particles by solidified Na_2S_4 at gaps between flat growing faces. Particles are covered with a thin layer of solidified Na_2S_4 with microclusters of other by-products.

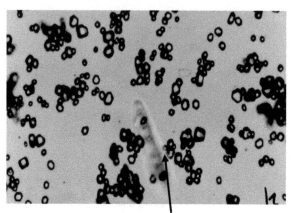

crystallized by-product

FIGURE 6.13
Optical microscope image (300X) showing large Na_2S_4 crystal in a commercial ZnS phosphor powder.

According to chemical handbooks, the solubility of ZnS in water is negligibly small (7×10^{-2} g per 100 g water) and Na_2S_4 is soluble in water [123]. Owing to the large difference in solubilities, it is believed that Na_2S_4 is easily removed from the produced ZnS powder by washing with water.

The complete removal of solidified Na_2S_4 from the phosphor powder is, however, a difficult task, especially from the gaps between flat faces. After washing several times with water, the completion of removal is usually monitored by measuring the electric conductivity of the rinse water, or by the addition of drops of $Ag(NO_3)$ solution to the rinse water to detect the presence of Cl^- in the rinse water. The rinsed phosphor powder dries in a heated oven at 110°C. Contrary to expectation, the by-products are not completely removed from the rinsed phosphor powder. The sintering material is Na_2S_4, and detection of Cl^- does not monitor the residual Na_2S_4. In the drying process, some amount of residual Na_2S_4 between the gaps dissolves in the heating water and then crystallizes to a large size from the condensed solution. Consequently, commercial phosphor powders usually contain crystallized by-products. Figure 6.13 shows a microscope image (300X) of a large Na_2S_4 crystal in dried phosphor powder.

The incomplete removal of Na_2S_4 (and by-products) generates problems, especially in the preparation of the screening slurry and storage of the phosphor powder. The problems relate to the (1) instability of the pH value and conductivity of the PVA phosphor slurry, (2) clumped particles of stored phosphors, (3) defect-laden phosphor screens (pinholes and poor edges), and (4) photodarkening of ZnS phosphor powders under irradiation by UV light and sunlight.

The results in Figure 6.13 suggest the complete removal of by-products from a commercial phosphor in 1996 [126]. The removal procedure involved washing the commercial phosphor powder in water agitated and heated overnight at 80°C. Particles of the washed phosphor powder had a chemically clean surface. Since then, the adhesion of microclusters (e.g., surface treatments) is unnecessary for CL phosphor powder. The phosphor powder takes away the troubles associated with clumped particles in the storage and instability of PVA phosphor slurries. Since 1996, the image quality on phosphor screens in color CRTs has markedly improved with the application of well-rinsed phosphor powders, without pigmentation. In many cases, however, the cleaned phosphor powders are deliberately contaminated with pigmentation on request from CRT producers, generating the flicker and smear images observed on many phosphor screens.

The removal of by-products by heated water is costly for phosphor production. The problems are caused by the melted by-products at 950°C. If the phosphor powders are produced without melted by-products in the heating mixture, and if the produced phosphor powders do not contain plate particles and particles smaller than 1 μm, the produced phosphor particles in the crucible are softly bound to solidified $ZnCl_2$ (2×10^{-4} mole), which has served as the flux. Because the pore sizes in the blend mixture in the crucible are controlled in a narrow range, the produced phosphor powder does not contain large plate particles. This is the case for the recipe shown in Table 6.8. Figure 6.14(A) shows an SEM micrograph (2000X) of the as-grown ZnS:Ag:Cl

(A)

(B)

FIGURE 6.14

SEM images (2000X) of ZnS:Ag:Cl blue phosphor particles in which $ZnCl_2$ is produced from raw $MgCl_2$; (A) as-produced particles that have MgS microclusters on the surface of particles, and (B) cleaned ZnS particles etched by dilute HCl solution.

phosphor particles produced with the recipe in Table 6.8. Phosphor particles as grown adhere to a small amount of microclusters of by-products [MgS ($T_m >$ 2000°C)] of raw material of the flux. The microclusters are smoothly removed by heated water and/or etched away with a dilute acid solution as seen in Figure 6.14(B). The ZnS phosphor powder shown in Figure 6.14(B) has no trouble in screening of PVA phosphor slurry without the aging process in preparation of the slurry of PVA phosphor. The phosphor powders do not clump during storage. Furthermore, the phosphor powder does not darken under exposure of the UV light and sunlight, even under wet conditions, indicating that the darkening of the ZnS phosphor particles relates to the Na_2S_4 residual. This is a significant improvement in ZnS phosphor production. The phosphor powders are produced in illuminated rooms without protection from UV light by placing yellow film on windows and illumination lamps.

6.8   Production of Practical ZnS Phosphor Powders

The fundamentals involved in ZnS phosphor production have been described. Now we delve into the practical production of highly optimized phosphor powders — in particular, screenable phosphor powders. The major concerns in the production of screenable phosphor powders include (1) no clumps of primary particles, (2) a narrow distribution of particle sizes, (3) uniform shape of primary particles, and (4) polycrystalline and micro-crystalline particles. This section describes the engineering details of how to produce practical phosphor powders.

6.8.1   The ZnS:Ag:Cl Blue Phosphor

The production technology of practical ZnS:Ag:Cl blue phosphor powders was established by Leverentz prior to 1945 [7]. Employing the recipe in Table 6.1, the produced phosphor powders emit a brilliant blue CL. The difficulties encountered in blue ZnS phosphor production, which many pro-ducers still encounter, include the (1) reproducibility of the CL intensity and color, (2) stability in the PVA phosphor slurry, and (3) screenability of the powders on the faceplate of CRTs.

The difficulties arise from (1) mishandling the quartz crucibles (in many cases, 500 ml) that charge the blend mixture, (2) the seal of the crucible with a lightweight coverplate, (3) misidentifying the flux material, (4) the role of the melted and vaporized sulfur for the elimination of oxygen contamination, (5) the growing space of ZnS particles in the pores (air bubbles), (6) overlooking the role of melted Na_2S_4 as the sealant, and (7) the removal of sintered by-products in the contact gaps between ZnS particles. Those problems are solved by (1) using tightly sealed crucibles with a heavy coverplate, (2) using the appropriate amount of raw material for $ZnCl_2$ flux (4×10^{-4} mol per mole a-ZnS powder), (3) using the appropriate amount of sulfur (3×10^{-3} mole per mole a-ZnS powder) in the blend mixture, (4) using a suitable/appropriate heating program for chemical reactions in heating crucibles, and (5) soaking the produced ZnS:Ag:Cl phosphors in water heated to 80°C for a time is sufficient for water to diffuse into contact gaps between particles.

A simplified and advanced production process is given below; a suitable recipe is given in Table 6.8. After the mixture of the recipe given in Table 6.8 is blended in a V-shaped mill, the blend mixture charges in a 3-L alumina crucible of the shape shown in Figure 6.8. One can use a quartz crucible, which adds to the production cost. An alumina plate, which has a slightly smaller diameter than the inner diameter of the crucible, is placed on the charged mixture. The crucible is placed in a furnace at RT, and the furnace heats according to the heat program shown in Figure 6.10. With large and tightly sealed crucibles, and the appropriate amount of halide and sulfur in the blend mixture, one can heat the crucible in an ordinary furnace that has a door. It is not necessary to put charcoal powder on the blend mixture, nor is it necessary to provide a flow of N_2 gas in the furnace during the heating process.

Now for the details of the chemical reactions and growth mechanisms. $MgCl_2$ reacts with melted S to form MgS, thereby releasing Cl_2 gas. The released Cl_2 gas reacts with ZnS to form $ZnCl_2$ ($T_m = 283$°C). When the crucible is maintained at 400°C, the coarsely blended mixture equilibrates with the melted S and by-products in the heating mixture. One can obtain the same quality of ZnS phosphor powders using different blending times (e.g., for 30, 90, or 120 min). At 400°C, the pores in the mixture are softly sealed with the melted mixture of S and $ZnCl_2$. The oxygen in the sealed pores reacts with evaporated S to form SO_2 gas. Then the temperature rises to 600°C at a heating rate of 400°C/90 min (= 4.4°C/min). During this process, the excess S evaporates above 445°C. The evaporated S reacts with oxygen in the mixture, and the oxygen is completely eliminated from the blend mixture in the crucible by conversion to harmless SO_2 gas. Using the slow heating rate, the expanding N_2 and SO_2 gases in the pores slowly leak into the crucible space. The furnace temperature is maintained for 1 hour, during which time the $ZnCl_2$ uniformly spreads in the mixture with low viscosity. The temperature of the crucible rises to 970°C at a rate of 270°C/ 60 min (= 4.1°C/min). During this temperature increase from 732°C, the growth of ZnS particles starts in the pores via chlorine cycles. The heating temperature of 970°C is determined by the desired particle size. The furnace temperature holds at 970°C for 60 min. The growth of ZnS particles may stop for 45 min at the temperature of equilibrium. Make sure that temperature equilibrium occurs at the center volume in the mixture. Then, the furnace turns off to cool down the crucible to RT. At 750°C, the $ZnCl_2$ vapor liquefies. The surface of the grown particles is covered by a thin layer of liquefied $ZnCl_2$, and the liquefied $ZnCl_2$ preferentially penetrates the contact gaps between ZnS particles. The $ZnCl_2$ thin layer solidifies at 283°C. Because the concentration of $ZnCl_2$ is very low, the grown ZnS:Ag:Cl phosphor

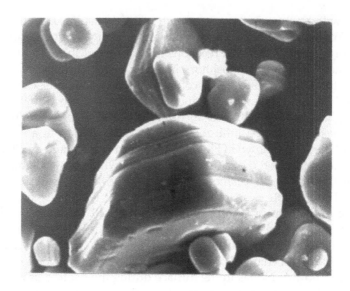

FIGURE 6.15
SEM micrograph of ZnS:Ag:Cl phosphor particles grown in a furnace that has ripples in the power supply (e.g., on-off regulation).

particles are softly sintered in the cooled crucible. This procedure results in the particles shown in Figure 6.14(A).

If control of the furnace temperature reveals the presence of ripples due to the use of a temperature regulator with poor accuracy, the shape of the grown particles is not smooth. Figure 6.15 shows, as an example, grown ZnS:Ag:Cl particles under poor temperature regulation. It is recommended to use a rippleless regulator for phosphor production.

The softly sintered particles are soaked in gently stirring water for at least 60 min. The wet ball-milling process is unnecessary in production. Then, the powder is lightly etched with a dilute HCl solution to remove the interface layer of ZnS and $ZnCl_2$. After confirming that the electric conductivity of the rinse water is less than 5 μS, the phosphor powder separates from the rinse water by filtration. The water should be removed from the filtrated powder as rapidly as possible. The filtrated phosphor powder dries overnight in a heated oven at 105°C. The dried phosphor powder sieves with a 400 mesh. The sieved powder is ZnS:Ag:Cl blue phosphor powder, which is now ready for the preparation of a PVA phosphor slurry. An example ZnS:Ag:Cl blue phosphor powder is shown in Figure 6.14(B).

6.8.2   ZnS:Ag:Al Blue Phosphor

Although ZnS:Ag:Cl phosphor powders are suitable as CL phosphors, some customers want to have the ZnS:Ag:Al phosphor. This requirement is nonsense, but phosphor producers respond to such requests. One difficulty encountered

TABLE 6.9

Recipe for Blend Mixture of ZnS:Ag:Cl Blue Phosphor

Material	a-ZnS	Ag	$Al_2(SO_4)_3$	ZnI_2	S	H_2O
Weight	3000 g	330 mg	150 mg	18 g	60 g	30 g

in the production of the ZnS:Ag:Al phosphor is the oxidization of the Al compound in the heating mixture. Table 6.9 gives the recipe for the ZnS:Ag:Al phosphor. The blend mixture heats with the heating program shown in Figure 6.11.

Another difficulty encountered in the production of ZnS:Ag:Al phosphors is the decomposition of ZnI_2 at temperatures above 624°C. ZnS particles no longer grow at temperatures above 750°C. The growth of the ZnS:Ag:Al phosphor particles should be controlled in the narrow temperature range between 500 and 750°C. The particle sizes in the sealed crucible change with the concentration of ZnI_2 in the blend mixture. Figure 6.16(A) shows, as an example, the average particle sizes as a function of the concentration of ZnI_2 in the blend mixture. With a constant heating rate (2.8°C/min.) between 500 and 750°C, the particle sizes increase with increasing ZnI_2 concentration. Figure 6.16(B) shows that the size decreases with heating rate. The ZnI_2 vapor can escape from the heating mixture using a fast heating rate. Practical phosphor powders (average particle sizes around 4 μm) will grow with ZnI_2 concentrations of around 2 to 4×10^{-4} mole fraction per mole a-ZnS.

The ZnS:Ag:Al phosphor powder that is grown at 700°C has a brown body color and emits poor CL. To bleach the color, the grown ZnS:Ag:Al phosphor powder is annealed at high temperatures (e.g., at 970°C for more than 60 min). The annealed phosphor powder is bleached and emits a brilliant blue CL.

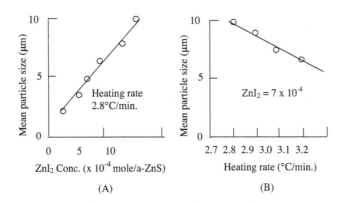

FIGURE 6.16

Average particle sizes of ZnS:Ag:Al phosphor as a function of added ZnI_2 concentrations with a constant heating rate (2.8°C/min) (A), and with a heating rate between 500 and 750°C at constant ZnI_2 concentration.

TABLE 6.10

Recipe for Blend Mixture of ZnS:Cu:Al Green Phosphor

Materials	a-ZnS	Cu	$Al_2(SO_4)_3$	ZnI_2	S	H_2O
Weight	3000 g	330 mg	150 mg	1.8 g	6 g	30 g

The conditions used for soaking, rinsing, and drying the heated product are the same as for ZnS:Ag:Cl phosphors.

6.8.3 The ZnS:Cu:Al Green Phosphor

The production of ZnS:Cu:Al green phosphors is essentially the same as the production of ZnS:Ag:Al blue phosphors, except for the Cu activator. The recipe for the blend mixture of ZnS:Cu:Al green phosphors is given in Table 6.10.

Polycrystalline ZnS blue and green particles, which distribute in narrow sizes with similar shapes, are produced by heating the blend mixture in large, tightly sealed crucibles. Microcrystalline ZnS phosphors having flat growing faces that grow with the addition of seed particles to the blend mixture.

6.9 Production of Y_2O_2S Phosphor Powders

The red primary in CL color display devices is the Y_2O_2S:Eu phosphor. The monochrome screen in a high-resolution CRT is the Y_2O_2S:Tb:Eu white phosphor. Y_2O_2S is a common host crystal, and activators include Eu^{3+} (red) and Tb^{3+} (green). Because Eu^{3+} (0.95 Å) and Tb^{3+} (0.92 Å) substitute for Y^{3+} (0.89 Å) lattice sites, these activators are stable in the Y_2O_2S crystal, with a negligible distortion of the Y_2O_2S lattice.

The raw materials are Y_2O_2, S, and I-a carbonates. A 4N purity of the raw materials is pure enough for the production of practical CL and PL. In the production of Y_2O_2S phosphors, the luminescence color is solely determined by Eu^{3+} (or Tb^{3+}) concentration, and the amount of Eu^{3+} (or Tb^{3+}) in Y_2O_2S phosphor particles determines the luminescence intensity. The variation in CL intensity in the produced phosphors derives from the contamination by oxygen during phosphor production. Eu^{3+} (or Tb^{3+}) is a multivalent element, and the valence smoothly changes oxidization state in the production process of Y_2O_2S phosphors. Eu^{3+} (or Tb^{3+}), which has changed valence via oxidization, does not form luminescence centers, which apparently reduces the activator concentration. The CL intensities of Y_2O_2S phosphors decrease with the degree of oxygen contamination during the heating process of phosphors, and the CL color shifts to shorter wavelengths as a result of the oxidization. Oxygen should be completely eliminated from the heating mixture before the formation of Y_2O_2S particles, similar to the production of ZnS phosphors, for quality control of the produced phosphor powders.

As discussed for the production of ZnS phosphors, the major source of oxygen contamination is air bubbles in the powdered blend mixture charged in crucible. Control of the furnace atmosphere and excess S powder in the charged mixture in the crucible does not have any effect on the elimination of oxygen in the charged mixture. There are two ways to eliminate oxygen in the heating mixture: one involves melting and evaporating S in the charged mixture, and the other involves an element that smoothly oxidizes in the heating mixture.

There is a report that the addition of a small amount (~100 ppm) of Tb (or Pr) to the blend mixture enhances the CL intensity by a few percent. In some cases, the Eu^{3+} CL intensity is enhanced by the addition of Tb^{3+}, and in other cases there is no enhancement in CL intensity. Careful experimental observations reveal the following. If the crucible is placed directly in the heated furnace at above 500°C, and is then heated to the temperature required for growth of the particles, the Eu^{3+} CL intensity is certainly enhanced by the addition of Tb. When the same crucible slowly heats from RT to 400°C, according to a proper heating program (and similar to the program for ZnS phosphor production), no change in CL intensity is observed with the addition of Tb. This result shows that oxygen in the blend mixture preferentially reacts with melted and evaporated S before reacting with Tb^{3+}. The measurements in the report are precise, but the samples were prepared using improper conditions. The claims of phosphor preparation in many reports are based on empirical results using improper heating programs, and the claims could not be verified using proper heating programs for the blend mixture in a crucible. The heating program of the blend mixture in a crucible is an important concern for the production of brighter phosphors. No complication (no in-house secret) will arise in the production of phosphors with the proper heating programs for the blend mixture.

We now discuss the flux materials for growth of Y_2O_2S particles. Y_2O_2S is an artificial compound produced by the heating of a mixture of Y_2O_3 and a large amount of S and Na_2CO_3. Table 6.11 gives the recipe for the Y_2O_2S:Eu phosphor [29, 30]. Chemical reactions in the heating mixture in a loosely sealed crucible include:

$$Y_2O_3 + Na_2CO_3 + 5S \rightarrow Y_2O_2S + xNa_2S_4 + (1 - x)Na_2S + CO_2(\uparrow) + O_2(\uparrow)$$

$$(6.4)$$

The by-products are Y_2O_2S, Na_2S_4 and Na_2S. When the blend mixture is heated in a large crucible with a tight seal provided by a heavy cover, only

TABLE 6.11

Recipe for Y_2O_2S:Eu Red Phosphor

Raw Material	Weight (g)	Molar Ratio
Y_2O_3:Eu (4.2 wt. %)	100	1.00
Na_2CO_3	45	1.00
Sulfur	71	5.0

Na_2S_4 is formed in the crucible, and Na_2S is not formed in the blend mixture. We can study whether the by-products of Na_2S_4 and Na_2S might add flux action to the growth of Y_2O_2S particles.

Melted Na_2S_4 is chemically reactive with SiO_2 (i.e., the quartz crucible). Only alumina (Al_2O_3) crucibles should be used for heating the blend mixture. In the early days of Y_2O_2S:Eu red phosphors per the recipe in Table 6.11, the crucibles, which had a volume of 500 ml, were placed directly into a heated furnace at temperatures above 500°C, and rapidly heated to around 1100°C. The produced Y_2O_2S powders usually had a light-brown body color. Pure Y_2O_2S particles are white. The brown body color resulted from the inclusion of Na_2S_4 in the particles [5]. In heating the blend mixture, the conversion of Y_2O_3 to Y_2O_2S particles occurs at around 700°C with the help of melted Na_2S_4. Raw Y_2O_3 particles are usually precipitated particles from the solution, and they are not well-crystallized particles. They are amorphous and contain many gaps. When the crucible rapidly heats to 1100°C, the melted Na_2S_4 (yellow) is of high viscosity and does not move out from the gaps of Y_2O_3 particles, and Y_2O_3 particles convert to Y_2O_2S seed particles. Consequently, the Y_2O_2S seed particles include Na_2S_4 (yellow) inside the particles, giving rise to a yellow body color.

The blend mixture slowly heats from RT to 400°C, and subsequently heats up to 700°C. The crucible is maintained at 700°C for 60 min; the conversion slowly occurs in blend mixture. The melted Na_2S_4 moves out from the gaps of Y_2O_3 particles, and then the conversion occurs in the heating blend mixture, resulting in a colorless Y_2O_2S powder. The heat program of the blend mixture between RT and 700°C is an important consideration in the production of colorless Y_2O_2S:Eu red phosphors. Figure 6.17 shows the heat program

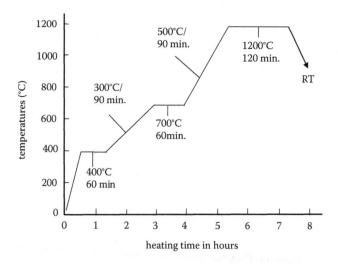

FIGURE 6.17
Heat program of furnace for production of Y_2O_2S:Eu phosphors in tightly sealed 3-L crucible.

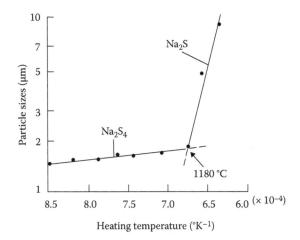

FIGURE 6.18
Particle sizes (average) of Y_2O_2S:Eu red phosphor produced by a mixture of Y_2O_3, Na_2CO_3, and S as a function of heating temperatures between 900 and 1300°C.

of Y_2O_2S:Eu red phosphors. When the particle sizes are determined by weight equivalent sizes, the size of the Y_2O_3 particle duplicates the size of the converted Y_2O_2S particle. The converted particles (1.5 μm) are seed particles [5].

When the crucibles heat up to various temperatures above 900°C, Y_2O_2S seed particles grow from the seed particles. Figure 6.18 shows the average particle sizes of Y_2O_2S as a function of heating temperature (K^{-1}) of the crucible between 900°C and 1350°C. The curve consists of two straight lines that inflect at 1180°C, which coincides with the $T_m = 1180°C$ of Na_2S. At temperatures below 1180°C, particle sizes are constant (= seed particles), indicating no flux action of the melted Na_2S_4. The melted Na_2S_4 merely acts as the conversion agent from Y_2O_3 to Y_2O_2S. Particle size sharply increases with temperature above 1180°C, indicating the flux action of the melted Na_2S. If the crucible is tightly sealed by the cover at temperatures above 1000°C, the heating mixture of Table 6.11 only contains Na_2S_4, and one does not obtain the curve shown in Figure 6.18. Formation of Na_2S in the crucibles is essential for the growth of Y_2O_2S. For the formation of Na_2S, some S should escape from the heating mixture. When a 500-ml crucible with a lightweight cover is used, and the crucible rapidly heats to above 500°C, a large amount of S in the blend mixture escapes from the crucible. The sealing conditions of the crucible and the amount of S in the blend mixture sensitively influence the formation of Na_2S in the heating mixture. Furthermore, the blend mixture heats up to 1250°C for the growth of practical Y_2O_2S particle size (4 μm). This is too high a temperature in production facilities. The cost of furnace bricks sharply increases above 1200°C. Therefore, the recipe given in Table 6.11 is not the optimal recipe for the production of Y_2O_2S phosphors.

The results depicted in Figure 6.18 suggest that melted I-a monosulfides may have flux action to Y_2O_2S particles. Table 6.12 shows the T_m values of

TABLE 6.12

T_m of I-a Sulfides

Sulfide	T_m (°C)
Li_2S	900–975
Na_2S	1180
K_2S	840

I-a sulfides. When the blend mixture contains some amount of Li_2CO_3 (Table 6.13), the growth curve of Y_2O_2S particles differs from Figure 6.18. Figure 6.19 shows the average particle sizes grown at temperatures between 800 and 1250°C. The particles are grown at temperatures between 900 and 1180°C, indicating that melted Li_2S (T_m = 900 to 975°C) has flux action to the growth of Y_2O_2S particles. Above 1180°C, the slope of the curve is the same as that in Figure 6.18, showing that the particles are grown due to melted Na_2S. Using melted Li_2S, particles of 4 μm are produced by heating at 1150°C, down 100°from 1250°C, although the crucibles are tightly sealed with the cover. The addition of a small amount of Li_3PO_4 helps the crystallization. With the addition of K_2CO_3 to the mixture, the effort is not clear in the curve as to whether melted K_2S has flux action. Table 6.13 provides a typical recipe for the production of Y_2O_2S red phosphors — as determined by a trial-and-error approach.

The next concern involves particle shape. Fortunately, the mixtures in Tables 6.11 and 6.14 contain large amounts of S (70 wt. % Y_2O_3). If the softly charged mixture heats from RT to 400°C and remains at that temperature for 60 min, as for ZnS phosphors, the charged mixture in the crucible decreases in volume, showing the equalization of pores (air bubbles) in the mixture. The particles are grown with the aid of a melted mixture of Na_2S_4 and Li_2S, which moves along the temperature gradient in the mixture. The particles no longer grow during heating for 150 min with 3-liter crucibles, indicating that the melted mixture of Na_2S_4 and Li_2S does not move in the mixture at an equivalent temperature. The grown particles distribute in a narrow size range and with similar shape. Figure 6.20 shows SEM images (5000X) of Y_2O_2S particles as grown for 120 min.

Plate particles are also grown in the densely charged mixture if the mixture of raw materials contains seed Y_2O_2S particles (1.5 μm). The amounts and

TABLE 6.13

Blend Mixture of Y_2O_2S:Eu
Phosphors with Li_2CO_3

Raw Material	Weight (g)
Y_2O_3:Eu (2×10^{-2} mole)	100
Na_2CO_3	70
Li_2CO_3	14
Sulfur	70

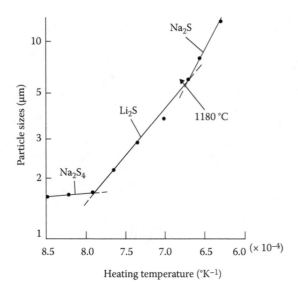

FIGURE 6.19
Particle sizes (average) of Y_2O_2S:Eu red phosphor produced by a mixture of Y_2O_3, Na_2CO_3, Li_3PO_4, and S as a function of heating temperatures between 800 and 1300°C.

sizes of plate particles in the produced Y_2O_2S phosphor powder are controlled by the addition of seed particles in the starting blend mixture.

Grown Y_2O_2S particles are minimally sintered in the crucible with a large amount of solidified Na_2S_4. The minimally sintered products in the crucible have a large heat capacitance, as compared with that of the floor (or carrying plate) in the furnace. To avoid breaking the bottom of the crucible because of a large temperature difference (thermal shock) between the crucible and the carrying plate (furnace floor), the contact area of the crucible should be minimized. Large crucibles that are tall and have a narrow diameter are preferred for the production of Y_2O_2S phosphor powder. Figure 6.21 illustrates a suitable crucible shape of 3 liters for the production of Y_2O_2S particles.

TABLE 6.14

Blend Mixture for the Production
of Y_2O_2S:Eu Phosphors

Raw Material	Weight (g)
Y_2O_3:Eu (2×10^{-2} mole)	100
Na_2CO_3	70
Li_2CO_3	15
Li_3PO_4	5
Sulfur	70

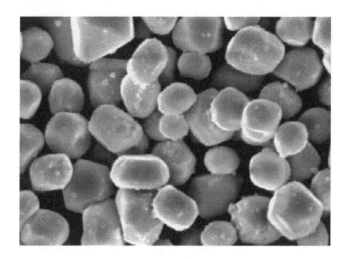

FIGURE 6.20
SEM image (5000X) of Y_2O_2S:Eu red phosphor produced by heating at 1050°C for 200 min.

Heated products in crucibles are minimally sintered with the large amount of solidified Na_2S_4 and Li_2S. The sintered by-products are well crystallized materials at high temperatures, which slowly dissolve in water even though the solubility given in chemistry handbooks show a high solubility. The Y_2O_2S phosphor particle is insoluble in water, and is negligibly small in both acid and basic solution. Therefore, the heated products in the crucibles are rinsed in water overnight for dissolution of the by-products. City water can be used for this procedure. Stirring and heating (70°C) the rinse water may help dissolve the by-products. The particles are separated to primary particles in the rinse water. Ball-milling of the rinsed powder is unnecessary.

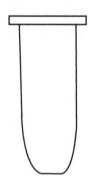

FIGURE 6.21
Suitable crucible shape for Y_2O_2S phosphor powder production.

Y_2O_2S particles are chemically stable materials in acid and basic solution. Therefore, the washed Y_2O_2S particles are etched by acid solution (5 to 10% HNO_3 solution) to remove the residuals of the by-products and the interface layer of Na_2S_4 and Y_2O_2S. After the etched particles are well rinsed with DI water, the powder dries in a heated oven at 110°C. The dried powder is Y_2O_2S phosphor powder for CL devices. Each phosphor particle has a clean surface, and the PVA phosphor slurry is stable with time.

7

The Design of Screenable Phosphor Powders for Cathodoluminescent Devices

It was shown in previous chapters that the cathodoluminescent (CL) properties of phosphor powders have undergone optimization for more than 35 years because the transition probability of luminescence centers is governed by the short-range perfection around these luminescence centers. Small variations in CL properties result from oxygen contamination. Powdered mixtures charged in crucibles inevitably contain oxygen at 3×10^{-3} mole per mole a-ZnS powder. Because the blend mixture contains sulfur powder of more than 3×10^{-3} mole per mole a-ZnS powder, the oxygen is completely eliminated by heating the mixture using a heat program prior to crystallization of the phosphor particles. There is little room for further improvement in the CL properties of phosphor powders. The remaining problem concerns the screening of phosphor powders onto the CRT faceplate. Screenable phosphor powders are composed of polycrystalline particles 4 μm in size, rather than spherical [127–133] and microcrystalline particles. Spherical particles exhibit poor adhesion to the faceplate, and the particles easily move from the position on the substrate by moving water.

Another real problem in the design of the screenable phosphor powders is the practical evaluation of the phosphor powders. The acceptability (screenability) of CL phosphor powders is made by CRT production engineers. At present, CRT production predominantly resides in Asian countries. CRT production engineers evaluate CL phosphor powders using existing screening facilities, which have been adjusted for the defects in phosphor powders. Naturally, the facilities do not accept an improved phosphor powder. CRT production engineers strongly control screening facilities with their empirical skills, and it is difficult to change their operating conditions. By considering the reality of CRT production, there is no generalized CL phosphor powder. We must produce three different kinds of phosphor powders to respond to the existing high demand and for development in the future. They are the phosphor powders for (1) ideal CL phosphor screen by improved screening technology, and (2) thick and (3) thin screens for established screen facilities.

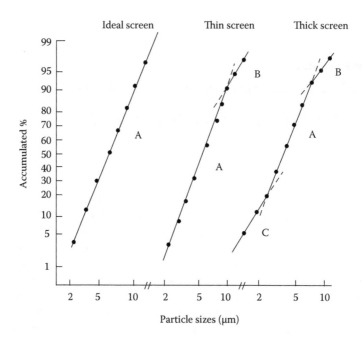

FIGURE 7.1
Three particle size distributions of phosphor powder for three different screening facilities: (A) is for ideal screen facilities. For established screening facilities, the phosphor powder must contain anchor particles (B) for thin screens; and for thick screens, the powder must contain both anchor (B) and small particles (C).

CL phosphor powders should not contain by-product residuals. As the phosphor powders are produced by the established production process, the produced phosphor powders are soaked in heated (80°C) and stirring water for 10 hours, in order to dissolve the by-products in the contact gaps between particles. When the produced phosphor powders do not contain the residuals of the by-products, the distribution of phosphor particles can be considered for screenable phosphor powders.

Figure 7.1 shows the distribution of the three categories of phosphor particles: (1) the ideal screen with new screening facilities, and the (2) thin and (3) thick screens for the established screening facilities. For phosphor powders that are properly produced in tightly sealed crucibles, and the crucibles in the furnace heat according to proper heating programming, the produced phosphor powder consists of primary particles (polycrystalline particles; A). The particles distribute with log-normal probability. The ideal screen is produced with only primary particles. In established screening facilities, the wet screen always dries before the particles settle on the substrate. As the screen thickness is less than three layers of particles, the powder must contain an appropriate amount of anchor particles (B), which quickly settle in the slurry and then strongly adhere to the wet substrate after moving a short distance. The anchor particles are plate particles distributed in large sizes of distribution A.

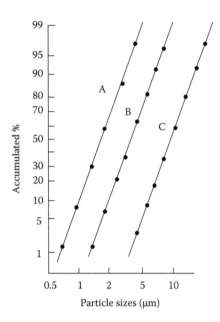

FIGURE 7.2
Particle size distribution of the ZnS:Ag:Cl phosphor with heating at temperatures of 830°C (A), 950°C (B), and 1030°C (C) for 90 min.

The maximum size of plate particles is 2 times the median particle size (at 50% of distribution) of A distribution. Screens thicker than five layers need some number of small particles (C) that generate crater pinholes in dried PVA phosphor screens. Without crater pinholes, the thick screen drops away from the substrate during the development process.

Because the appropriate amount of B and C change with individual screening facilities in CRT production, the acceptable phosphor powder for them is made by a mechanical blend of the appropriate amounts of A, B, and C, responding to individual screening conditions. Therefore, it is recommended that anchor B and small C particles are separately produced. The anchor particles (microcrystals in plate shape) are grown in the crucible if the blend mixture contains seed particles (~1 µm). The small particles are grown at a low heating temperature. Figure 7.2 shows the distribution of ZnS particles grown by heating at 830°C (A), 950°C (B), and 1030°C (C) for 90 min. The distribution of the particle sizes on log-normal graph paper shifts in parallel with the heating temperature. Similar results are obtained with Y_2O_2S phosphor powder.

8

Conclusion

The phosphor screens in cathodoluminescent (CL) display devices have provided light images since the cathode ray tube (CRT) was invented in 1897. CRTs are well-established display devices that serve as an interface between the human eye and electronics devices that record information from modern life activities. This is because image pixels in a phosphor screen emit approximately 10^{21} photons s^{-1} cm^{-2}, which is comparable to the number of daytime scenes that human eyes have adjusted to for 7 million years. The large number of photons from CL phosphors is due to the high energy conversion efficiencies of CL phosphors (20% for ZnS), which were optimized in the 1970s both theoretically and practically. Furthermore, CRTs display the images on phosphor screens with low power consumption (30 W per total screen area) using a scanning electron beam of high density. That low power consumption is a great advantage over LCDs and PDPs. Power consumption of the LCD backlight would be 350 W to obtain comparable image luminance. The images on phosphor screens are acceptable to viewers who watch the images at a distance of 5 times the screen size.

Since the 1970s, many CRT manufacturers have moved out of the United States and Europe to Japan, and also to South Korea, China, and other Asian countries, countries that focus primarily on production costs. In these countries, production engineers have a major say in decisions regarding the acceptance or rejection of particular phosphor powders. Improved phosphor powders are solely evaluated by the screenability of the phosphor powder using a particular production facility. In such evaluations, the image qualities of phosphor screens in CRTs reflect the condition of the phosphor screen with all its defects. Thus, the produced CRTs display images that exhibit moiré, flicker, smear, a low contrast ratio, and poor color fidelity, which are perceived at the distance of distinct vision (30 cm away from the screen). This situation therefore allows challenge from FPDs (such as LCDs and PDPs), which display flicker-free images on the screens, although FPDs are power-hungry display devices with low energy conversion efficiencies from input power to visible light (about 1/20 that of CRTs).

The image quality of phosphor screens in CRTs can be improved by improving the quality of the phosphor powder. Unfortunately, the ambiguities concerning the characteristic properties of phosphors, the phosphor powder production process, and the screenability of phosphor powders have remained for more than 50 years. We have clarified these ambiguities in this book. CL properties generated in phosphor particles are *quantitatively* — not qualitatively — explained using our knowledge of solid-state physics. The energy conversion efficiencies of CL phosphors were optimized empirically and theoretically in the 1970s; that is, state of the art. The small variations in CL intensities and CL color in phosphor production are due to contamination by oxygen (3×10^{-3} mole per mole a-ZnS powder), which is inevitably contained in powdered blend mixtures charged in crucibles. The large amounts of oxygen are completely eliminated from the heating mixture before crystallization of phosphor particles, by the control of heating programs of the furnace and by the use of the proper crucibles. The by-products melted at high temperature preferentially penetrate the gaps between grown particles, and grown particles bind due to the solidified by-products at room temperature. The bound particles in the produced phosphor powders generate the difficulties encountered in the screening of phosphor powders on the faceplate of CRTs. Using the by-products that have a melting temperature higher than 1000°C, there are no bound phosphor particles in the heated crucibles. Incorporating knowledge of the chemical reactions in the heating mixture and the behavior of by-products in the heating crucible, the produced phosphor powders do not contain the residuals of by-products; and the phosphor particles have a clean surface. The produced phosphor powder consists of primary particles of the same shape, and the particles distribute in a log-normal manner with a small standard deviation.

The flicker of images on CRT phosphor screens is due to perturbation of the electron trajectory of incoming electrons by negative charges (surface-bound electrons, SBEs) in front of insulators, which adhere to phosphor particles. The surface of commercial phosphor powder particles is heavily contaminated with deliberately adhered microclusters of SiO_2 (surface treatment), pigments, and by-product residuals of phosphor production. The contaminants are insulators, and the SBEs on contaminants effectively shield the phosphor particles negatively. To obtain flicker-free images on phosphor screens, the contaminants should first be completely eliminated from the surface of phosphor particles (i.e., clean surface). Then, the phosphor screen must be constructed with two layers of phosphor particles on a conductive substrate, including the black matrix. With such improved screens, the negative field of the SBEs on the top layer of the phosphor screen is concealed by the anode field, which allows penetration of incoming electrons into the phosphor particles without perturbation, resulting in flicker-free images on the phosphor screen. The moiré is also eliminated from the phosphor screen.

Smear, poor contrast ratio, and whitening of brighter images on the phosphor screen are caused by spreading of scattered CL lights to neighbor phosphor pixels in different colors. Because practical phosphor particles have a high index of refraction, comparable to that of diamond, emitted CL light is widely scattered by reflection on the surface of phosphor particles in the screens.

The scattering of CL light can be confined in each phosphor pixel in the phosphor screens by surrounding the phosphor pixels with a barrier that absorbs the light. Then, CL phosphor screens can display images that are comparable with the printed images on sheets of paper.

Another important concern of display devices is the lifetime of the devices. The lifetime of CL display devices is determined by the vacuum technology, not by the CL phosphors. By application of the improved vacuum technologies, the phosphor screens may hold original luminance for the lifetime of the heater in BaO cathodes (longer than 10 years). On the other hand, the lifetime of LCDs is determined by the lifetime of the backlight, wherein the luminance exponentially decreases with operating time. The nominal lifetimes of FPDs are estimated at half the initial luminance. FPDs have a short lifetime compared to the lifetime of CRTs. We will see an increase in the practical lifetime of FPDs in the future.

The improved phosphor powders still encounter difficulties in the evaluation of screenability by established screening facilities, which produce phosphor screens with defects. The present phosphor screens are produced utilizing the through-pinholes to the faceplate. Control of these defects limits any improvement in phosphor screens. The defect-laden phosphor screens do not allow for a scientific study of photolithography. With a good comprehension of the established screening facilities, one can produce screenable phosphor powders by mechanically mixing the primary particles and using appropriate amounts of anchor particles and small particles that have clean surfaces. The phosphor screens may display improved images even though the phosphor screen is not the best screen.

As a long-range prospect, CL devices possess the superiority of image luminance, image quality, long lifetime, low production cost, and energy savings over LCDs and PDPs, even though the device is bulky and heavy. The high image quality protects the human eye from permanent damage, and the long lifetime and low production cost are of benefit to the consumer's economy and against environmental pollution. These are the trade-offs of today's bulky and heavy CRTs. Thus it can be predicted that CL display devices will always have a marketplace niche.

Luminescence centers of PL phosphors in illumination lamps are excited directly via the charge-transfer bands by absorption of UV light from a 253-nm mercury (Hg) lamp. The penetration depth of the incident UV light into individual phosphor particles is about 0.1 μm, but incident UV light can also penetrate into the deep layers of phosphor particles, which were laid down on phosphor screens, by light scattering on the surface of phosphor particles. The number of photons absorbed by individual phosphor particles is small compared with the number of luminescence centers existent in phosphor particles. The luminance of PL devices is linearly related to the exciting UV intensity. One can obtain 5 to 7 times the luminance with the same PL lamp if the exciting UV light intensity is increased. This means that PL phosphor screens maintain high energy conversion efficiencies by extended exciting light intensities. Consequently, the brighter PL illumination lamps will be obtained by increasing the UV light intensity to a higher level than present lamps.

This is a very important consideration of development fluorescent lamp and plasma display devices, especially development of a practical Hg-free fluorescent lamp. The diameter of the corona discharge of Xe and Hg + Ar gases in discharge chamber is influenced by the strength of the negative field of SBEs which tightly bind with the induced holes in phosphor particles in the screen. The removal or minimization of SBEs on phosphor screens in the discharge chamber gives a narrow gap between the discharge and phosphor screen, resulting in the increase in the UV lights on phosphor screens in the discharge chamber, by reduction of self-absorption of the emitted UV lights. This problem can be solved by a comprehension of physical properties of the bulk of the phosphor particles, which I have described in Sections 4.4 and 4.5.

A good comprehension of generation mechanisms of CL and PL, excitation of phosphor particles in screens, and preparation of highly optimized phosphor powder, as well as physical properties of bulk of phosphor particles, will contribute to an improvement in the standard of living.

References

1. K.F. Braun, *Ann. Phys. Chem.*, 60, 552, 1897.
2. H.S. Nalwa and L.S. Rohwer, Eds., *Handbook of Luminescence, Display Materials and Devices*, Vols. 1–3, American Scientific Pub., Stevenson Ranch, CA, 2003.
3. L. Ozawa, *Application of Cathodoluminescence to Display Devices*, Kodansha, Japan, 1993, pp. 313–318.
4. L. Ozawa and M. Itoh, *Chem. Rev.*, 103, 2836–2855, 2003.
5. L. Ozawa, *Cathodoluminescence, Theory and Application*, Kodansha, Japan, 1990, pp. 237–253.
6. F.A. Kröger, *Some Aspects of the Luminescence of Solids*, Elsevier, New York, 1948.
7. H.W. Leverentz, *An Introduction to Luminescence of Solids*, John Wiley & Sons, New York, 1950.
8. D. Curie and G.F.J. Garlick, *Lumincescence in Crystals*, Methuen & Co. Ltd., London, 1960.
9. M. Balkanski and F. Gans, in *Luminescence of Organic and Inorganic Materials*, H.P. Kallman and G.M. Spruch, Eds., John Wiley & Sons, New York, 1962, p. 318.
10. S. Shionoya, in *Luminescence of Inorganic Solids*, P. Goldberg, Ed., Academic Press, New York, 1966, chap. 4.
11. G.F.J. Garlick, in *Luminescence of Inorganic Solids*, P. Goldberg, Ed., Academic Press, New York, 1966, chap. 12.
12. I. Broser, in *Physics and Chemistry of II–VI Compounds*, M. Aven and J.S. Prener, Eds., North-Holland, Amsterdam, 1967, p. 526.
13. C.S. Scott and C.E. Reed, Eds., *Surface Physics of Phosphors and Semiconductors*, Academic Press, New York, 1968.
14. G. Blasse and B.C. Grabmaier, *Luminescent Materials*, Springer-Verlag, Berlin, 1994.
15. S. Shionoya and W.M. Yen, Eds., *Phosphor Handbook*, CRC Press, Boca Raton, FL, 1998.
16. P.N. Yocom, *J. Soc. Inf. Display*, 4, 149, 1996.
17. M. Yamamoto, *J. Soc. Inf. Display*, 4, 165, 1996.
18. *Electrochem. Soc. Interface*, 12(2), "Luminescence and Display Materials," 2003.
19. L. Ozawa, M. Makimura, and M. Itoh, *Mater. Chem. Phys.*, 93, 481, 2005.
20. L. Ozawa and M. Itoh, *Semiconductor FPD World*, 7/04, 114, 2004, in Japanese.
21. L. Ozawa, *Mate. Chem. Phys.*, 73, 144, 2002.
22. L. Ozawa and S. Hayakawa, *J. Luminesc.*, 8, 325, 1983.
23. L. Ozawa and K. Oki, *Mat. Chem. Phys.*, 60, 274, 1999.
24. L. Ozawa, U.S. Patent 6,825,604 B2 (2004).
25. L. Ozawa, M. Koike, and M. Itoh, *Mat. Chem. Phys.*, 93, 420, 2005.
26. L. Ozawa, M. Makimura, and M. Itoh, *Mater. Chem. Phys.*, 96, 511, 2006.
27. D.J. Robbins, B. Cockyane, J.L. Glasper, and B. Lent, *J. Electrochem. Soc.*, 126, 1222, 1979.

28. H. Yamamoto and T. Kano, *J. Electrochem. Soc.*, 126, 305, 1979.
29. M.R. Royce, U.S. Patent 3,418,246,12968.
30. P.N. Yocom, U.S. Patent 3,418,247,12968.
31. L. Ozawa, *J. Electrochem. Soc.*, 128, 140, 1981.
32. G.W. Ludwig and J.D. Kingsely, *J. Electrochem. Soc.*, 117, 348, 1970.
33. J.D. Kingsely and G.W. Ludwig, *J. Electrochem. Soc.*, 117, 353, 1970.
34. M. Godlewski, K. Swiatek, A. Suchocki, and J.M. Langer, *J. Luminesc.*, 48/49, 23, 1991.
35. A. Rothwarf, *J. Appl. Phys.*, 44, 752, 1973.
36. H. Yamamoto and A. Tonomura, *J. Luminesc.*, 12/13, 947, 1976.
37. R.C. Alig and S. Bloom, *J. Electrochem. Soc.*, 124, 1136, 1977.
38. K. Era, S. Shionoya, Y. Washizawa, and H. Ohmatsu, *J. Phys. Chem. Solids*, 29, 1843, 1968.
39. A. Suzuki and S. Shionoya, *J. Phys. Soc. Jpn.*, 31, 1455, 1971.
40. Yui-Shin Fran and Tseung Tseng, *J. Phys. D; Appl. Phys.*, 32, 513, 1999.
41. P.J. Dean, *Luminescence of Inorganic Solids*, P. Goldberg, Ed., Academic Press, New York, 1966.
42. A. Suzuki and S. Shionoya, *J. Phys. Soc. Japan*, 31, 1455, 1971.
43. M. Ikeda, L. Ozawa, K. Hata, Z. Shinozuka, and G. Ishizuka, *Radioisotopes*, 13, 115, 1964.
44. L. Ozawa and H.N. Hersh, *Phys. Rev. Lett.*, 36, 683. 1976.
45. L. Ozawa, *J. Electrochem. Soc.*, 129, 1535, 1982.
46. R.N. Bhargava, D. Gallengher, X. Hang, and A. Nurmikko. *Phy. Rev. Lett.*, 72, 416, 1994.
47. K. Sooklal, B.S. Cullum, S.M. Angel, and C.J. Murphy, *J. Phys. Chem.*, 100, 4551, 1996.
48. J.P. Yang, S.B. Qadri, and B.R. Ranta, *J. Phys. Chem.*, 100, 17255, 1996.
49. J.P. Yang, H.F. Gray, D.S.Y. Hsu, S.B. Qadri, G. Rubin, B.R. Ranta, W.L. Warren, and C.H. Seager, *J. Soc. Inf. Display*, 6, 139, 1998.
50. M. Tanaka, S. Sawai, M. Sengoku, M. Kato, and Y. Masumoto, *J. Appl. Phys.*, 87, 8535, 2000.
51. M.H. Lee, S.G. Oh, S.C. Yi, D.S. Seo, J.P. Hong, C.O. Kim, Y.K. Yoo, and J.S. Yoo, *J. Electrochem. Soc.*, 147, 3139, 2000.
52. M. Ihara, T. Igarashi, T. Kusunoki, and K. Ohno, *J. Electrochem. Soc.*, 149, H72, 2002.
53. K. Manzoor, S.R. Vadera, N. Kumar, and T.R.N. Kutty, *Mater. Chem. Phys.*, 82, 718, 2003.
54. T. Phan, M. Phan, N. Vu, T. Anh, and S. Yu, 2004, *Phys. Stat. Sol. (A)*, 201, 2170, 2004.
55. J.S. Kim, H.E. Kim, H.L. Park, and G.C. Kim, *Solid State Commun.*, 132, 459, 2004.
56. Sekita, K. Iwanaga, T. Hamasuna, S. Mohri, M. Uota, M. Yada, and T. Kijima, *Phys. Stat. Sol. (B)*, 241, R71, 2004.
57. G.H. Li, F.H. Su, B.S. Ma, K. Ding, S.J. Xu, and W. Chen, *Phys. Stat. Sol. (B)*, 241, 3248, 2004.
58. C. Ye, X. Fang, G. Li, and L. Zhang, *Appl. Phys. Lett.*, 85, 3035, 2004.
59. E. Martinez-Sanchez, M. Garcia-Hipolito, J. Guzman, F. Ramos-Brito, J. Santoyo-Salazar, R. Martinez-Martinez, O. Alvarez-Fregoso, M.I. Ramos-Cortes, J.J. Mendez-Delgado, and C. Falcony, *Phys. Stat. Sol. (A)*, 202, 102, 2005.
60. H.C. Swart, A.P. Greeff, P.H. Holloway, and G.L.P. Berning, *Appl. Surf. Sci.*, 140, 63, 1999.

61. L. Ozawa and M. Itoh, *Semiconductor FPD World*, 9/04, 100, 2004, in Japanese.
62. S. Kubonia, H. Kawai, and T. Hoshina, *Jpn. J. Appl. Phys.*, 19, 1647, 1980.
63. R. Raie, M. Shiiki, H. Matsukiyo et al. *J. Appl. Phys.*, 75, 481, 1994.
64. R. Raue, M. Shiiki, H. Matsukiyo, H. Toyama, and H. Yamamoto, *J. Appl. Phys.*, 75, 481, 1994.
65. A. Bril and H. A. Klasens, *Philips Res. Rep.*, 7, 401, 1952.
66. T. Hase, T. Kano, E. Nakazawa, and H. Yamamoto, *Advances in Electronics and Electron Physics*, Academic Press, New York, Vol. 79, p. 311, 1990.
67. V.D. Meyer, *J. Electrochem. Soc.*, 119, 920, 1972.
68. S. Shionoya, T. Koda, K. Era, and H. Fujiwara, *J. Phys. Soc. Japan*, 19, 1157, 1964.
69. L. E. Shea, *Interface*, 7, 24, 1998.
70. M.G. Craford and G.B. Stringfellow, Eds., *High Brightness Light Emitting Diodes*, Academic Press, San Diego, 1997.
71. S. Watanabe, *OyoButsuri*, 74, 1437, 2005, in Japanese.
72. L. Oosthuizen, H.C. Swart, P.E. Viljon, P.H. Holloway, and G.L.P.Berning, *Appl. Surf. Sci.*, 120, 9, 1997.
73. D.C. Lee, S.A. Bukesov, S. Lee, J.H. Kang, D.Y. Jeon, D.H. Park, and J.Y. Kim, *J. Electrochem. Soc.*, 151, H227, 2004.
74. L. Ozawa, *Display and Imaging*, 5, 151, 1997, in Japanese.
75. L. Ozawa, *Display and Imaging*, 5, 59, 1996, in Japanese.
76. G. Hplle and C. Curtin, *Information Display*, 4&5/00, 34, 2000.
77. H. Bechtel, W. Czarojan, M. Haase, and D. Wadow, *J. SID*, 4/3, 219, 1996.
78. H.C. Swart, J.S. Sebastian, T.A. Trottier, S.L. Jones, and P.H. Holloway, *J. Vac. Sci. Technol. A*, 14, 1697, 1996.
79. L. Oosthuizen, H.C. Swart, P.E. Viljoen, P.H. Holloway, and G.L.P. Berning, *Appl. Surf. Sci.*, 120, 9, 1997.
80. H.C. Swart, A.P. Greeff, P.H. Holloway, and G.L.P. Berning, *Appl. Surf. Sci.*, 140, 63, 1999.
81. H.C. Swart, *Phys. State. Sol. (C)*, 1, 2354, 2004.
82. S. Itoh, T. Kimizuka, and T. Tonegawa, *J. Electrochem. Soc.*, 136, 1819, 1989.
83. K. Oki and L. Ozawa, *T. IEE, Japan*, 114-C, 26, 1994.
84. T. Nakamura and K. Kiyozumi, Japan Patent 45-20223, 1970.
85. T. Nakamura and K. Kiyozumi, VFD and its applications, Nikkan Kogyo Shinbunn, Tokyo, Japan, 1977.
86. T. Kishino, Ed., *VFD*, Sangyo Tosho, 1990, in Japanese.
87. L. Ozawa and M. Itoh, *Semiconductor FPD World*, 8/04. 102, 2004, in Japanese.
88. S. Itoh, T. Tonegawa, T.L. Pykosz, K. Morimoto, and H. Kukimoto, *J. Electrochem. Soc.*, 134, 3178, 1987.
89. H. Hiraki, A. Kagami, T. Hase, K. Mimura, and Y. Narita, *1975 International Conference on Luminescence*, Tokyo, Japan, P208, Session BO7-4, 1975.
90. E. Miyazaki and Y. Sakamoto, *Proc. SID*, 29, 19, 1988.
91. M. Watanabe, K. Nonomiya, M. Ueda, T. Kanehisa, Y. Moriyama, and J. Nishida, *SID '85 Digest*, 1985, p. 185.
92. M. Watanabe, *Proc. SID.*, 29, 187, 1988.
93. T. Utsumi, *IEEE Trans. Electron Devices*, 38, 2276, 1991.
94. R.H. Flower and L. Nordheim, *Proc. Roy. Soc. London, Ser. A*, 119, 173, 1928.
95. C.A. Spindt, *J. Appl. Phys.*, 39. 3504, 1968.
96. H.F. Gray, G. J. Campisi, U.S. Patent, 4,964,946, 1990.
97. C.A. Spindt, C.E. Holland et al., *IEEE Trans. Electron Dev.*, ED 38, 2355, 1991.
98. L. Ozawa, *Display and Imaging*, 5, 143, 1997.

99. R. O. Petersen, *Information Display*, 3/97, 22, 1997.

100. E. Yamaguchi, K. Sakai, I. Nomura, T. Ono, M. Yamanobe, N. Abe, T. Hara, K. Hatanaka, Y. Osada, H. Yamamoto, and T. Nagagiri, *Proceedings of IDW'97*, 1997, pp. 52–55.

101. T. Kusunoki, M. Sagawa, M. Suzuki, A. Ishizaka, and K. Tsuji, *Proc. IDW'00*, 2000, pp. 956–962.

102. A. Hiraki, T. Ito, and A. Hatta, *Oyo Butsuri*, 66, 235, 1997.

103. S. Uemura, T. Nagasako, J. Yotani, T. Shimojo, and Y. Saito, *Tech. Digest of SID 98*, 1998, pp. 1052–1055.

104. T. Komada, Y. Hionda, T. Hatai, Y. Watabe, T. Ichihara, K. Aizawa, Y. Kondo, and K. Koshida, *Tech. Digest of SID 00*, 2000, pp. 428–431.

105. L. Ozawa, *Materials Chemistry and Physics*, 51, 107, 1997.

106. R.A. Pearson, *Information Display*, 7, 8, 22, 1991.

107. C.A. Maxwell, *Information Display*, 11, 7, 1992.

108. American National Standard for Human Factors Engineering of Visual Display Terminal Workstations, American National Standard Institute, ANS/HFS Standard No. 100-1988.

109. H. Bruning, *Physics and Application of Secondary Electron Emission*, Pergamon, London, 1954.

110. J.J. Quinn, *Phys. Rev.*, 126, 1453, 1962.

111. M.W. Cole and M.H. Cohen, *Phys. Rev. Lett.*, 23, 1234, 1969.

112. W.A. White, *Naval Res. Rept.*, 1950, p. 3666.

113. K. Ohno and T. Kusunoki, *Digest Tech. Paper SID*, 2004, p. 1048.

114. J. Kosar, *Light Sensitive Systems*, John Wiley & Sons, New York, 1965, chap. 2.

115. P.B. Brown and W.H. Fonger, *J. Electrochem. Soc.*, 122, 94, 1975.

116. W.H. Fonger, *Appl. Opt.*, 21, 1219, 1982.

117. T. Oguchi and M. Tamatani, *J. Electrochem. Soc.*, 133, 841, 1986.

118. W. Busselt and R. Raue, *J. Electrochem. Soc.*, 135, 764, 1988.

119. K. Oki and L. Ozawa, *J. Soc. Inf. Display*, 3/2, 51, 1995.

120. L. Ozawa and X. Li, *J. Soc. Inf. Display*, 6, 285, 1998.

121. L. Ozawa, M. Makimura, and M. Itoh, *Mater. Chem. Phys.*, 93, 481, 2005.

122. H. Kawai, T. Abe, and T. Hoshina, *Jpn. J. Appl. Phys.*, 20, 313, 1981.

123. D.R. Lide, Ed., *CRC Handbook of Chemistry and Physics*, 74th ed., CRC Press, Boca Raton, FL, 1993.

124. E.G. Zubler and F. A. Mosby, *Illum. Engs.*, 54, 734, 1959.

125. J.W. van Tijen, *Philips Technical Rev.*, 23, 237, 1961/1962.

126. L. Ozawa, *Display and Imaging*, 4, 111, 1996, in Japanese.

127. N. Yoshida and K. Mikio, U.S. Patent 5,413,736, 1995.

128. Y. Pei and X. Liu, *Chinese J. Lumin.*, 17, 5, 1996.

129. Y.D. Jiang, Z.L. Wang, F. Zhang, H. Paics, and C.J. Summers, *J. Mater. Res.*, 13, 2950, 1998.

130. S.H. Cho, J.S. Yoo, and J.D. Lee, *J. Electrochem. Soc.*, 145, 3, 1998.

131. A, Veht, C. Gibbons, D. Davies, X. Jing, P. Marsh, T. Ireland, J. Silver, and A. Newport, *J. Vac. Sci. Technol. B*, 17, 750, 1999.

132. X. Jing, T. Ireland, C. Gibbons, D.J. Barber, J. Silver, A. Veht, G. Fern, P. Trown, and D.C. Morton, *J. Electrochem. Soc.*, 146, 4654, 1999.

133. S.H. Cho, S.H. Kwon, J.W. Oh, J.D. Lee, K.J. Hong, and S.J. Kwon, *J. Electrochem. Soc.*, 147, 3143, 2000.

Index

A

Aizawa, K., 59
American National Standard Institute, 65
Angel, S.M., 28, 38, 60
Aven, M., 3, 10

B

Balkanski, M., 3, 10
Berning, G.L.P., 31, 38, 50
Bhargava, R.N., 28, 38, 60
Braun, Karl Ferdinand, invention of cathode
ray tube, 1
Busselt, W., 81, 87, 89
By-products
physical properties, 112
removal of, 122–126

C

Campisi, G.J., 58
Cathodoluminescence, photoluminescence
generation, phosphor particle,
143–146, overview
faceplates, screening of phosphor powders
on, 81–102
image quality improvement, phosphor
screens, 63–80
luminance improvement, display device
phosphor screen, 41–62
phosphor particles, luminescent property
generation, 7–40
phosphor powder production,
103–138
screenable phosphor powder design,
cathodoluminescent devices,
139–142
Chen, W., 28
Cho, S.H., 139
Color CRT luminance improvements,
41–43
Color phosphor screens, 94–101

Crucible
sealing conditions, 115–116
size/shape, 119–120
ZnS particle growth in, 113–120
Cullum, B.S., 28, 38, 60

D

Design, screenable phosphor powders,
139–142
Diamond, refraction index, 12
Display device screen luminance, 41–62
color CRT luminance improvements, 41–43
decrease, CL intensity during operation,
50–51
field emission display, 58–61
flat CL displays, 52–61
horizontal address vertical deflection, 56
human eye perception, image luminance
on screens, 45
lifetime, phosphor screens, 46–52
modulation deflection screen, 57
oxide cathode lifetime, 51–52
phosphor pixel operation conditions, field
emission display, 60
photoluminescence device lifetime, 52
power consumption, 53
residual gases, coloration of phosphor
screens, 46–50
spot luminance, screen luminance,
correlation, 43–44
vacuum fluorescent display, 53–56
Drying unwedged phosphor screen, 84–85

E

Electric properties, phosphor particles, 69–76
Energy conversion efficiencies, phosphor
particles, 37–39
Era, K., 11, 16, 37
Excitation mechanisms, luminescence
centers, phosphor particles, 7–11
Eye perception, image luminance on screens,
45

Related Titles

Organic Electroluminescence, Zakya Kafafi
ISBN: 0824759060

Inorganic Phosphors: Compositions, Preparation and Optical Properties, William Yen
ISBN: 0849319498

Phosphor Handbook, Second Edition, William Yen
ISBN: 0849335647

Milton Keynes UK
Ingram Content Group UK Ltd.
UKHW040052071024
449327UK00019B/507